One

ALBERT'S BLIND SPOT

The ascertainment by Michelson and Morley in 1887 of an accurate measure of the speed at which light normally travels is undoubtedly one of science's major milestones. It made possible new and precise calculations in researches into the nature of electromagnetic energy. Also, it alerted Albert Einstein, among others, to the possibility that weird things would be seen by observers of objects at great distance and observers of objects travelling at high speed. Einstein's mathematics told him that Isaac Newton's concept of *absolute* time was mistaken, that is, rather than absolute, that the time associated with every point in space is unique, that the time associated with an object travelling at high speed would be passing more slowly than normal and that the mass of an object would increase as the speed of that object increased. These revolutionary ideas of Einstein's were published in 1905 in what has come to be known as the Special theory of relativity. Alas, Einstein's reasoning was flawed. He had misunderstood and misinterpreted the results of his calculations.

Taking account of the speed of light, the Special theory points to the fact that two observers at different distances from a distant event will see that event at different times. And, very roughly, this disagreement on time forms the basis of the postulation that there is no *absolute* time, and that distance (or space) and time are inextricably bound. In other words, as the distance between the two observers increases the observed time discrepancy increases correspondingly.

Other examples of the theory focus on what is observed when bodies are moving. As the distance between an observer and a distant event increases the recorded time of the event will be increasingly delayed, i.e., time will be 'dilated' and clocks that are observed will appear to be slowing down. But, contrary to what the theory claims, that is all that they do - they just *appear* to slow down. This distinction Einstein did not understand. The knowledge he would have needed was simply not available to him. In his day it was believed that the eyes were simply windows on to the world and that sight provided us with an objective and virtually infallible representation of the world about us. Einstein and his contemporaries believed, as most scientists still believe today it seems, that we can only see an event *as it is actually happening.*

The times observed and recorded in the above examples are subjective, perceived times - the events at these points are purely psycho-physiological ones or, where recording devices are used, photographic/electronic ones. The times recorded by the observers are simply the times at which they perceive the distant event and, by the time they so perceive, things at the distant event have moved on. In these examples it is only at the distant point that a real event, that is, an event in the physical world has taken place.

Because light travels at a finite speed everything we see is, to some extent, in the past. We can never see anything as it actually is. Even when I look at my watch on my wrist I do not see the time it is actually displaying, that is, the actual position of the hands. There is a lapse of time between pulses of light being reflected from the face of my watch and their arriving at my eyes. So, by the time a pulse of light which corresponds to a particular time arrives at my eyes the hands on my watch have moved on slightly, albeit, in this case, infinitesimally.

We see by means of our brains receiving and interpreting the patterns, continuously forming on our retinas, of points of electromagnetic energy called photons. And from this input our visual system constructs a 'picture'. The pattern of light tells us the shape, texture and position of objects and the frequency and strength of its pulses gives rise to the creation, by our visual systems, of the qualities of colour and brightness. And, the image we have of the world around us, forming on our retinas, is updated by every pulse of light that is reflected from it.

Now, if I move 186,000 miles away from my watch the light that is reflected from its face will take one second to arrive at my eyes. In other words, my visual updates will be delayed by one second. So, if when I glance at my watch I see the hands at, say, 'on the hour', by the time the pattern of light that represents 'the hour' reaches me my watch has moved on one second. So, when I see the hands of my watch at 'on the hour' I know that, at that moment, they are actually at one second past. And, if I move further away from my watch my visual updates from it will take even longer to reach me and the 'picture' I have of my watch will be correspondingly more out of sync. But, no matter how far away I am from my watch, or how long it takes the light that is reflected from its face to reach me, and therefore how slow my watch appears to be, my watch is still ticking away at its normal speed and keeping good time, just as it was designed, built and regulated to do.

If I now start to move steadily away from my watch my visual updates from it will be steadily and increasingly delayed. And if I move very quickly my watch will appear to have almost stopped - as my updates take longer and longer to reach me. If I were to reach the speed of light I would get no new updates at all, as I would be retreating exactly as quickly as they were coming towards me, and my watch would appear to have stopped. But of course, throughout my hypothetical high-speed journey, again, my watch continues to tick away at its normal speed. In other words, the apparent slowing

3

down of my watch and its eventual standstill has been a visual illusion.

Such visual illusions are not caused by any variations in the speed of light. The speed of light remains constant. It is detectable changes in the visual information that it carries that arrive slower, and it is this variation that deceives our eyes.

The Special theory would have predicted the apparent slowing and eventual stopping of my watch. But Einstein did not appreciate that the slowing of my watch is simply a visual illusion.

And, of course, it does not matter whether it is me or my watch that speeds off into space. Either way all that is happening is that my visual system is being deceived. At any speed both my watch and I still tick over at our usual, designed speeds. Einsteinian relativity of time postulates, however, that in the circumstances described in my imaginary experiment my watch really does slow down.

And, of course, Einstein could only interpret this as evidence of Time itself slowing down. And if Time slows down so does everything else, including a person's ageing.

The night sky is full of ambiguity. Skilled observers know where this exists and understand also how and why our eyes can sometimes get things wrong. It seems that most astronomers and physicists know, for example, that the apparent simultaneous explosion of two stars at different distances from Earth is simply a visual illusion and that our eyes are being deceived, because the distances involved are such that the light from the more distant star which exploded long ago, and the light from the nearer star which exploded more recently, is arriving at Earth at the same time. However, when it comes to clocks on spacecraft, they seem to lose this insight, and visual illusion is ignored.

Two spacecraft, A and B, have been prepared for a mission. Before lift-off, the two spacecraft's clocks and the controller on

Earth's clock are synchronized. The ships leave planet Earth and eventually arrive at their respective destinations, planet A and planet B. Planet A is one light hour away from Earth (that is, the distance light travels in one hour). Planet B is two light hours away from Earth. The two planets are on roughly the same line of sight from Earth. Shortly after they have both landed the controller contacts Captain B and instructs him to come out of his craft the following day with his ship's signalling lantern and clock, to point these items towards Earth and to give one flash of light at exactly 4 o'clock. Captain A is told to carry out exactly the same procedure, but at 5 o'clock. Next day, as instructed, Captain B flashes his lamp once towards Earth at exactly 4 o'clock. At 5 o'clock Captain A does likewise. On Earth, as 6 o'clock approaches the controller aims his telescope towards the two planets. Suddenly, he sees B's and A's lights flash simultaneously. He checks his clock. It is exactly 6 o'clock. He looks at spacecraft B's clock. It says 4 o'clock. A's clock is at 5 o'clock. But this is precisely what he expected. He knows that the light from planet B would take two hours to reach Earth and that the light from planet A would take one hour to reach Earth, and that he would see the two lamps flash together. He knew too that the light that was reflected from B's clock when it was at 4 o'clock, and the light that was reflected from A's clock when it was at 5 o'clock, would also arrive at Earth at the same time. The controller seeing B's clock at 4 o'clock and A's clock at 5 o'clock is a visual illusion. And the controller knew that although he was seeing 4 o'clock and 5 o'clock, all three clocks were actually at 6 o'clock. In other words, because of the delay in the arrival at Earth of the light reflected from clocks A and B, the controller sees B's clock as it was two hours ago and A's clock as it was one hour ago. Einstein's calculations showed similar time discrepancies, but he misinterpreted this as a dilation of Time, rather than simply an increase in the time taken for light to span a particular distance.

Einstein, and it seems all of today's physicists and astronomers, would argue that, in the above example, when the observer saw the two spacecraft's clocks saying 4 o'clock and 5 o'clock it is because they *were* at 4 o'clock and 5 o'clock and that the two lights appeared to flash together because they did flash at the same time. They would so argue because the Special theory postulates that there is no 'absolute' time and that the time at which an event takes place is determined by the distance from which the event is observed. This postulation is based, it seems, on the belief that if we see something happening it is because it really *is* happening. But this reasoning is based on a misunderstanding of the mechanics of seeing. Because of the finiteness of the speed of light we never see things as they are, we are always seeing things as they were. The further away things are, the greater is this discrepancy.

If our spaceship A lifts off again, and begins to travel into space away from Earth and our observer on Earth occasionally checks the time on the spacecraft's clock he will see that the clock is apparently slowing down. But, just as before, he is not surprised. This is what he was expecting. He knows that this apparent slowing of the spacecraft's clock is due to the increasing delay in the light that is reflected from the clock's face arriving at his eyes, and he knows that if the spacecraft now accelerates the clock will appear to slow down at an increasing rate. But the observer knows that this is effectively a visual illusion, caused by the fact that the light that is reflected from the clock is taking longer and longer to arrive at Earth. Einstein, however, argues that the spacecraft's clock appears to slow down because it *is* slowing down. He postulates that in space, Time slows down due to movement through space and that in the case of accelerated movement Time slows down at an increasing rate. People, therefore, Einstein and his supporters would argue, age more slowly in space than they do on Earth.

The 'Twin Paradox', a well-established and generally accepted part of Modern Physics theory states that a twin who has travelled in space will return to Earth younger than the twin who has remained on Earth. An interesting, if not disturbing thing happens if we project the Special theory so that our spacecraft reaches the speed of light. At the speed of light not only will the spacecraft's clock appear to have stopped, since at this speed no new reflections would be leaving its face. But the hearts of the spacecraft's crew would appear also to have stopped. Would they be dead? But this mistaken postulation is also due to a lack of understanding of visual perception. We see events and objects when the light that is reflected from them arrives at our eyes. The more this light is delayed, the more we see things as they were and the slower events seem to be taking place. Things seem to be happening more slowly as they move away from us and, just as we might rationally expect, they appear to speed up as they move towards us.

If I observe a man walking on a distant planet he will appear to be at, lets say, point A. However, because of the great distance involved and the finiteness of the speed of light, I know he is actually at B some point further along the road. If I now fly down to meet up with the man, at point C, observing him as I go, I will see him walking more quickly than normal (how much more quickly depending on how quickly I travel) because *visually* he has had to walk from A to C in the same time it took him to *physically* walk from B to C. But, of course, he is not *actually* walking quicker - he just *appears to me* to be. The 'fast-forward' version of the man's journey has occurred only in my head through my brain receiving and interpreting the faster than normal arrival of visual information about him.

Einstein ascertained that the finiteness of the speed of light created a gap between the occurrence of an event and our observation of it. He bridged this gap by postulating that Time dilates. But this dilation occurs only inside our heads, in mathematical calculations and within recording instruments, and not in reality at a distant event. The formulae in question predict visual illusions. Clocks, watches, people and events do not actually slow down and distances do not really shrink, but they appear to do so to an observer because in one set of circumstances the visual information he is receiving is retarded and in the other it is advanced and his visual system is deceived accordingly.

And yet, in modern physics textbooks and countless books on space-time these illusions are presented as real - Time really does dilate and distances really do contract. They are often described as a consequence of the laws of nature, or the way the Universe works. So, how could this crazy state of affairs have come about? It has come about because in Einstein's day little was known about visual perception. Today, visual perception is well understood. However, because the answer to a problem in physics lies in psychology, and because neither of these two disciplines pays attention to what is happening in the other, Einsteinian ideas have gone largely unchallenged.

We now know that seeing is an active, cognitive process. And that although our brain gets it virtually right practically all the time, when it is presented with strange information, or when information is presented in a strange way, it can and will be fooled. Also, because light travels at a finite speed, there is always a discrepancy between the taking place of an event and our perceiving it - a lapse, corresponding and proportional to the distance involved. This causes what is also, effectively, a visual illusion - a phenomenon that Einstein unwittingly predicted in his Special theory.

Contrary to the Special theory, an event is not repeated each time it is observed. An event can take place only once. It is the seeing of an event that occurs again when it is observed from different distances. Seeing is entirely subjective. Where movement is involved, and when objects, such as spacecraft, are moving away from us or towards us, again, contrary to the Special theory, Time does not dilate or shrink, it simply appears to do so because of either advancement or delay in our receiving visual information from these objects. Time has no physical properties.

Einstein was also concerned with how mass might be affected by movement. He calculated that the mass of a body would increase as it began to move and, that the greater the velocity the greater would be the increase in mass. However, as common sense alone tells us, mass has nothing to do with movement and everything to do with a tot up of protons, neutrons and electrons. As ever, Einstein's maths is flawless. But, here, again, Einstein misinterpreted some of the consequences of the finiteness of the speed of electromagnetic radiations.

Let us imagine bouncing a radio signal off the tail and off the nose of an aeroplane. The plane is about a mile away on a runway and facing away. Although sent at exactly the same time, these signals will arrive back at the transmitter/receiver at slightly different times. This is simply because one has travelled further than the other. The return journey to the plane's nose is longer than the journey to and from the tail. A simple calculation involving these two times, the difference between them and the speed of the signals will give us the length of the aeroplane. Let us say that this works out at 50 feet. If we now imagine the plane in the air and flying away from us we will now find that the discrepancy between the two returning signals has increased. Based on these new times, a calculation of the plane's length could now work out at (say) 51 feet. If the plane now steadily increases in speed the return journey times of the two signals will increase accordingly suggesting that the aircraft is continuing to increase in size, and therefore, mass. However, this increase in mass is effectively an illusion. All that is happening is that, because the aircraft is now travelling at speed, the radio signal heading for the nose is virtually having to play catch-up. And, while it is doing so, the signal to the tail has already been reflected and is well on its way back to the transmitter/receiver. On the return journey, all recorded times get steadily smaller than before, suggesting that the plane is getting shorter and losing mass. Again, of course, this is not so. At no time throughout its flight does the aircraft gain or lose mass. The experiment has simply confirmed that radio signals travel at a finite speed and that this must always be taken into account in any calculations about mass and movement.

Two

CURVED SPACE

Curved space is the term most commonly used to describe the
explanation of gravity that Einstein offers in his General theory of
relativity. Not based on any experimentation or discovery, it is a
purely mathematical theory. It is a formula for the construction of
space that is contoured in such a way as to hold the Universe
together. There is little doubt that it is mathematically sound but as
well as being a contradiction in terms, curved space has difficulty
with the test that any theory must pass to qualify as a good theory -
the how-well-does-it-fit-reality test.

  The model most used to explain curved space comprises a
trampoline-like surface, a heavy sphere (such as a large ball
bearing) and a smaller, lighter sphere. When the heavy sphere is
placed on the trampoline it sinks into the elastic surface, curving the
material around it down towards its base. The small sphere is
pushed onto the apparatus in such a way that it orbits at a high level
on the sloping apparatus. As it loses momentum, its orbits will
become narrower and narrower and at a lower and lower level on
the slope until eventually it comes to rest against the large sphere. If
we released three small spheres onto the apparatus one after the
other, and we timed this well, we could have these three small
spheres orbiting simultaneously at different distances from the large
sphere. This is a fair model of our own solar system. Experts say
that Earth's orbital speed is very gradually reducing, that the
diameter of its orbit is getting gradually smaller and that ultimately
in a few billion years time the Earth will crash into the Sun. Is Earth
inching towards the Sun, as these experts tell us it is, because it is

on a slope? The model may be a good predictor of the pattern of life an orbiting planet will follow, but do the small spheres and planets share a similar fate for similar reasons? Are the planets of our solar system literally on a slippery slope towards their destruction?

During a solar eclipse in 1919 astronomers reported being able to see a distant star which they had calculated should have been obscured by the Sun. And, in agreement with Einstein's prediction, they concluded that the light from this distant star was being bent towards Earth by the curvature of the space around the Sun. This is perhaps the strongest evidence there is for the existence of curved space but it is by any standards flimsy. As they cannot be checked out, we have little choice but to accept that the calculations which put the distant star behind the Sun and out of sight were correct. Also, the reason why the distant star was visible remains a moot point and perhaps a more plausible explanation than curved space is that the light from the distant star was refracted by the Sun's atmosphere.

If curved space exists we should have to look no further than our own solar system to see evidence of it. If the Sun's mass is bending the space around it the orbital path of the planets of the solar system should be describing a saucer or shallow bowl shape around the Sun but this is not the case. Apart from one rogue planet, the planets of our solar system are orbiting on the same plane, and this is a flat plane which cuts through the Sun's core. Also, if we search the night sky we see no evidence of curved space. Saturn's rings are flat, and, rather than the bowl shape that curved space would produce, the galaxies too are remarkably flat.

The General theory of relativity is concerned with the way the Universe is held together. It offers an explanation as to the position of a planet in relation to other planets and stars but it makes no attempt to explain what is happening on the planet itself. On Earth, where it is not hampered or affected in any way by other forces or pressures, gravity is clearly trying to pull everything, including every molecule of water in the oceans, towards the Earth's core along the shortest, most direct route. Also, everywhere on Earth light and every other form of electromagnetic energy travels in straight lines. In the early days of radio and television, in order to get signals much beyond the horizon, we had to build relay stations. Today, to get radio and television signals around the globe, we bounce them off accurately positioned orbiting satellites. Where on Earth is there any evidence that space is curved?

One theorist argues that light travels too quickly to be caught in the curved space around Earth but this, rather than supporting the General theory, questions it. By definition, nothing can exist outside the contours of space. In the case of the star which should have been obscured being visible during the solar eclipse mentioned earlier, Einstein argues that it was not the light that was bending, it was the curved space within which it was travelling.

A curved space theory of gravity begs the question - why curves? Curves and slopes do not *cause* things to move, they *allow* things to move. Curves and slopes simply allow the pull of gravity to pull. Without gravity the orientation of surfaces is of no consequence. We know from actual experience that objects in outer space are weightless. If we were to place a heavy ball bearing on a trampoline in outer space it would not make the slightest dent. On Earth, the weight of objects that we experience is the physical expression of the force of gravity. We tend to think that it is the weight of an object that takes it down a slope. But, really, it is gravity that pulls it. In outer space, objects are weightless if and when they are too far away from the centre of a gravity field. Nowhere in Space can a slope or a curve make an object weighty. It is only the force of gravity that can give matter weight.

'Dimpled space', as it has come to be known, is a mathematical contrivance - a blueprint for a gravity simulator. It does not explain gravity. It is an as-if account of something that is probably beyond human understanding.

It is surely very unlikely that we will ever be able to explain a force whose job description includes pulling the tears we shed down our cheeks, keeping everything on Earth from floating off into Space and, at countless sites throughout the Universe, holding in a huddle a billion stars.

As we have seen, the force of gravity is a mystery. Over 300 years ago Isaac Newton ascertained how bodies behave when they are within the grasp of gravity. But nobody has ever been able to explain what gravity actually is. Gravity is generally thought of as lines of force, which consist of a constant stream of particles called gravitons which flow towards a gravitational source. But, this conventional view of gravity as gravitons does not make sense. Earth's gravitational field obviously extends at least as far as the Moon. So, if the Moon was being held in its orbit by lines of particles racing towards the Earth's core there would have to be a source of these somewhere beyond the Moon. And this is rather complicated by the fact that Earth is held within its orbit of the Sun by the Sun's gravitational pull. So, the Earth's gravitational field and the Sun's field must overlap somewhere in Space. And, if particles are involved this means that there would be a stream of particles from somewhere beyond the Moon flowing to Earth overlapping with a stream going from Earth towards the Sun. If gravity works as a stream of particles the Earth and the Moon would have to be encased in an atmospherical graviton factory that was able to provide an endless supply of gravitons. And, the entire solar system would have to be similarly encased.

If there are particles involved in the force of gravity it must be very doubtful as to whether they are constantly on the move. There is no obvious good reason as to why particles would have to be flowing in order to create attraction between two gravitational bodies. There is almost certainly no flow of particles involved in magnetism. It seems that there is a field of energy around each pole of a magnet and, if and when this field overlaps and interacts with another such field, depending on the polarities involved, attraction or repulsion immediately starts. There would appear to be a sphere of attraction around bodies of mass, such as Earth. And, rather like magnetism, it seems that if and when this field encounters a similar field surrounding another body a force of attraction is immediately set up. If unfettered, these two bodies would collide. However, it seems that throughout Space some force, such as momentum, acts to keep them apart.

On gravity, the General theory of relativity does little more than reveal Einstein's lack of appreciation of the problem that gravity poses. Very briefly, Einstein's gravity requires Newton's gravity to make it work. Einstein calculated that as a star forms it sinks into space creating a vast indentation around it. The mass of the star would determine how much it would sink and therefore how steep would be the slopes in the space surrounding it. If the path of a moving body such as a planet were to cross the slopes of curved space created by a star, for example, the smaller body, according to Einstein, would get entrapped and would go into orbit around that star. It is argued that if a star started collapsing and became more dense it would sink steadily deeper into space causing the slopes of space around it to get steeper and steeper. If this process were to continue and the star reached or got close to infinite density the indentation it had created would now be a hole and the sides of this would be vertical. And any body such as a star or another planet that virtually dropped into this hole would not be able to escape. Presumably this could happen if the hole was directly on the line of travel of the star or the planet. In fact, according to Einstein nothing would be able to escape from the enormous pulling force that would here exist - not even light. This is a black hole. But this is a very naive notion of gravity.

Einstein's explanation of the force of gravity is unconvincing. It begins by proposing that Space is curved. How curves can be formed out of emptiness or nothingness is not explained. However, according to Einstein, throughout the Universe, Space is curvy. These curves, it is claimed, are created as bodies of mass, stars, planets and moons, press down into Space. Here, Space has to be imagined as a sheet of space-time. The bigger and more massive the body the more it presses down into space-time and the steeper and more extensive are the slopes or curves it creates around itself. But, as we know, bodies cannot press downwards. There is no up or down in the Universe. Also, bodies of mass do not have weight and do not press in any direction. Rather, they are pulled and exert a force in a particular direction by being within the sphere of attractiveness of another body. So, if curves could be created in space, it would require Newtonian gravity to create them. And, if we have Newtonian gravity there is no need for curves.

If, on Earth, a ball is placed on a flat, horizontal surface it will remain quite stationary. If the flat surface is tilted slightly the ball will immediately roll downwards and at the edge of the surface drop straight to Earth. If the surface is tilted steeply the ball will roll much more quickly towards the edge. In a gravitational environment up is away from, and down is towards, the source of gravity. In a gravityless situation a ball placed on a flat surface would remain quite still as the surface was tilted. In fact, if the surface was removed altogether the ball would still not move - it would be suspended in space. And, so it is in outer space. If a planet was placed on a section of Einsteinian curved space it would not move. There are no forces compelling it to do so. In space there is no up or down, and the steepness of a slope is entirely of no consequence. If a body in motion were to travel into an area of Einsteinian curved space it would certainly follow that curve in one direction or another. But it would not accelerate according to the slope of that curve. It would need the force of gravity to make it do so.

Gravity, as curved space, means that everything that flows towards a star, for example, is diverted around that star. And studies have been, and are being, carried out to demonstrate this. But rather than lend support to the notion that the force of gravity is exerted by gradients such studies reveal its shortcomings. If everything that flowed towards stars and planets were diverted around them nothing would be able to land on their surfaces. A spacecraft returning from a mission to one of the planets or to the moon would start to dip as it approached Earth. Continuing to follow the curvature of space, it would dip down below Earth, go up and out the other side and on into space. Not even the rays of the sun would be able to land. And, if this were the case, and all radiations from the Sun were diverted around Earth the planet would be perpetually dark, ice bound and lifeless. Clearly, the pull of gravity does not rely on a flow of particles or curves. Gravity is a 24-carat conundrum that might never be solved.

The notion that the Universe is encased in a stretchy sheet of nothingness is, as it sounds, daft. And, without curvable emptiness Black Holes cannot be formed. Even if it were resting on a skin of bendable space, a star that is out on a limb, whether in a collapsed state or in its full glory, cannot actually do anything. Gravitationally, it takes two to tango. Every body has its own gravity, but it is only when, gravitationally, one body is within the sphere of attraction of another that things start to happen. In this situation, two bodies start pulling towards each other. And, the strength of this attraction is set by the mass of the two bodies and how far apart they are. Normally one body dominates the other, because it is greater in size, and this body's gravity is stronger. The mutual attraction between two bodies draws them together but just by an infinitesimal amount in a year. It has been calculated, for example, that Earth is creeping towards the Sun at perhaps just one or two inches a year.

The strength of the attraction between two bodies, due to their distance apart, therefore, is very stable. However, if the mass of one body or the other were to be altered considerably the gravitational attraction between the two would alter accordingly. The strength of the gravitational field associated with a collapsing star does not alter. This is because although its density is increasing, its mass is not changing. It is mass that determines gravity. So, unless there has been a dramatic change in a gravitational field somewhere, a collapsed star has not been pulled down a hole, or created very steep or even shallow slopes around itself onto which neighbouring stars or planets will be drawn. The claim that a newly collapsed star will suddenly start to suck in bodies that were previously well beyond its normal gravitational attraction is nonsense.

A small object will sink further into a soft surface than a large object of the same mass. This is because its downward force is concentrated into a smaller area than that of the larger object. But, of course, it is Earth's gravitational field that is giving both objects their downward force in the first place.

As the volume of a body decreases its density increases. Its mass and so its gravity, though, do not alter significantly. Of course, stars do lose mass continuously throughout their life. Despite their being no great loss of mass during a collapse, Black Hole supporters maintain that a collapsing star, overcoming electromagnetic repulsion and literally squeezing atoms together, continues to reduce in volume until its gravity becomes so strong that nothing can escape its clutches. Not even light can get away. Quite simply, this is wrong. Gravity can only act on mass, and light has no mass.

Some theorists argue that collapsing stars inevitably go on reducing until they reach a state of infinite density and no volume. This is preposterous. If it were the case the Universe today would be a completely empty place. Matter can not be destroyed, or created.

Nature is, as it has to be, the perfect recycler. Nothing is ever lost. We refer to stars as having a lifespan and talk about their death. But nothing really dies. In fact, stars get dismantled, moved on and reassembled elsewhere. The atoms and elements that comprise stars get badly buffeted all the time. They get fused, split, mixed and combined, but, some how, they get repaired, restored and returned to what and how they were originally - ready to become the building blocks of brand new stars. And, this recycling is perpetual. If it were not so we would not have the clearly defined and orderly periodic table we have today, and hydrogen would not be by far the most abundant element in the Universe.

Clearly, many scientists enjoy mathematically bending and stretching their minds, but putting forward theories, that have been worked out only on paper or computer, without being well supported by real evidence, is an extremely irresponsible thing to do. Crazy theoretical constructs like string theory, multiple universes, space-time, curved space, Black Holes and the Big Bang are not just distractions, albeit interesting ones, they positively hinder the advancement of scientific knowledge and understanding. Today, cosmological investigation is wandering aimlessly in a dense fog of spurious notions. A return to genuine progress is undoubtedly contingent upon getting back to sound methods involving exploration and discovery, and theories built only upon solid bits of real evidence. This could make development painfully slow but, if it does, so be it. If minds and methods do not alter there is unlikely to be any real progress made in cosmology in the foreseeable future.

Three

THE TROUBLE WITH HUBBLE

The most recent major assertion by astronomers and physicists about the Universe as a whole is that it is expanding. That is, that the galaxies are drawing apart. In other words, that the distances between the galaxies are increasing. And the evidence in support of this seems irrefutable since like almost everything we know about the Universe it is up there in the night sky for all, theoretically at least, to see. But things are not always what they seem.

The first step towards the picture of the Universe that is most widely supported today was made by B.M. Slipher in 1914 when he reported that the light from some distant galaxies he had observed was reddish. Inspired by this, Edwin Hubble began a search for distant galaxies. He eventually pinpointed more than thirty and discovered not only that all were reddish but also that the more distant the galaxy was, the redder it was. This 'red shift' told Hubble that every galaxy was moving away from our own and that the further away a galaxy was the further and faster it was moving.

A lowering of the frequency of light (i.e. the number of times it pulses per second) moves it towards the red end of the light spectrum. And this red shift can, according to some theorists, be caused by the movement of a source of light away from the point of observation. This 'Doppler Effect' is perhaps better known as an acoustic phenomenon, where the pitch of a sound is higher as its source approaches than it is when it passes and starts to go away.

Hubble concluded from his studies that the Universe was expanding in the sense that every galaxy was going away from every other galaxy, that the Universe has an outer region and that the galaxies are moving further and faster apart the nearer they are to this outer region. How can every galaxy be moving away from every other galaxy?

Gustav Tamman likens what the Universe is doing to what happens to a cake containing raisins while it is baking. Typically during baking, the outer area of the cake rises further and faster than the area nearer the centre, which means that raisins near the outer area will be pushed further apart than raisins nearer the centre. If the raisins are thought of as galaxies this model fits perfectly the Hubble picture of the Universe. But does Hubble's evidence fit the model?

The diagram at Figure 1 is a cross-section of a model of the Universe. Most astronomers would describe the Universe as finite, expanding and roughly spherical.

The larger dotted-line circle in the diagram indicates the outer reaches of the Universe and the smaller dotted-line circle indicates its inner edge. Few theorists speculate as to what there is or is not at the centre of the Universe but after several billion years of expansion from an earlier, smaller universe, we might reasonably expect that the Universe's core is roughly spherical and void. The solid line circle in the diagram indicates the extent of the visible Universe and the tiny circle at the centre of this, labelled E, represents our own galaxy. The solid line circle could, of course, be much smaller than is indicated, perhaps even as small as the tiny circle representing Earth's galaxy. However, since most astronomers would agree that even our largest telescopes cannot reach the outer edges of the entire Universe, the solid circle cannot be much larger than is suggested here.

The diagram at Figure 2 should be imagined as a small detail of a much larger diagram. In it there are three galaxies, one of which is our own. Galaxies X and Y are nearer the outer edge of the Universe than Earth's galaxy at E. We can assume that X and Y are

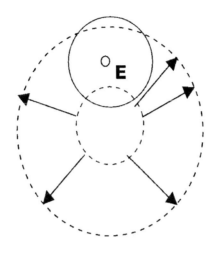

Figure 1

moving at the same speed, and that they will cover the same
distance in the same time. This is a configuration which could well
be repeated many times around our own galaxy. It may be helpful to
imagine the arrows in Figure 2 as following the path of the spokes
of a large wheel, and that the galaxies are drawing away from its
hub.

This means, therefore, as shown in Figure 2, that although Y
galaxy is much more distant from Earth than X, because it is
travelling on a radial path that is almost parallel to the path along
which Earth is travelling, its movement away from Earth is
relatively small. This is immediately at odds with Hubble. Hubble
maintains that the more distant a galaxy is, the more it is moving
away from Earth. And yet, if the Universe is expanding, providing
Earth is not exactly at its centre, it is almost inevitable that there
will be many galaxies moving on radial paths that are roughly
parallel to Earth's path and so will be moving away from Earth less

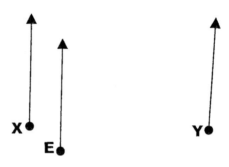

Figure 2

than many galaxies roughly on the same path as Earth. Also, if we consider that an observer on Earth will, following Hubble, see the most distant galaxies as the most reddish, we would expect Galaxy Y to be redder than X. However, according to red shift theory it is the galaxies that are moving away the most that appear the reddest.

So, although Y is further away than X, because very little of Ys movement is away from Earth, its colouration will appear to have shifted very little to the red. Galaxy X's movement, on the other hand, is almost entirely away from Earth. This will make it look redder than Y. Again we are at odds with Hubble who states that the *further* away a galaxy is the redder it appears. In our example, X is nearer to Earth than Y, and so should, according to Hubble, be less red than Y, which, as we have seen, it is not. Does this mean we can reject the Hubble studies? No. We can only partially reject his analysis and conclusion. We cannot reject what Hubble actually saw. He discovered that the further away a galaxy is, the redder it appears. Hubble interpreted this as indicating that the galaxies were receding. However, it may well be that his finding that the more distant a galaxy is the redder it is may be telling us no more than the observation itself suggests - that redness is a function of distance.

In the example above, there would be many galaxies behind and following Earth's galaxy. Let us say that one of these galaxies, Z, is much further away from E than X or Y. According to Hubble, because of the great distance involved, this galaxy should appear red. But is Hubble correct? The galaxy Z is moving towards E, but the distance between E and Z, in line with the raisin-cake model is increasing. This is because E, since it is nearer the outer edge of the universe, is travelling faster than Z. So, Z will not appear red - it is not moving away from E. Although the distance between E and Z is increasing, a redness is only seen if the distant galaxy is moving away from the point of view of an observer. There should be many such distant galaxies that do not appear red when observed from Earth.

It is most unlikely that the reddish appearance of distant galaxies has anything to do with movement. The sound waves from a moving source of sound such as the siren of an ambulance are pushed together by that movement. This raises the frequency of the sound waves, and this is why the pitch of a siren of an approaching ambulance sounds higher than normal. When the ambulance passes and draws away the pitch drops. This is known as the Doppler Effect. But a moving source of light like a star cannot similarly affect the radiations of its light. Light comprises straight-line pulses of electromagnetic energy travelling at 186,000 miles per second. This bears practically no comparison to sound waves which are ripples of molecules of air. Because of the immense speed of light its wavelength will be virtually unaffected by any movement of a source of light. Imagine a gang of bank robbers making a quick getaway in a car. One of the gang is hanging out of a window and shooting back towards a policeman who is shielding himself behind a post-box outside the bank. If the robber fires once per second when the car is stationary bullets will arrive at the policeman at the rate of one per second. Even at 60mph, as the robbers make their getaway, if the gun is still fired once per second the bullets will still arrive at the policeman at roughly one per second. This is because 60mph in one direction has little affect on the 1,000mph of the bullet in the opposite direction. At 60mph, the increase in *distance* that this would create between successive bullets would only be great enough to delay the bullets arriving at the policeman by a few hundredths of a second. This is an imperceptible change. As far as the policeman is concerned, throughout the robber's escape, bullets have continuously arrived at the post-box at about one per second. The pulses of light that come from a star are roughly analogous to the bullets that came from the bank robber's gun – they are little affected by any movement of the star. Again, this is largely due to the difference between the extremely high speed of light pulses coming from the star and possible speeds of the star itself.

Obviously, an ambulance can travel fast enough to change the pitch of the sound from its siren, but could galaxies be drawing away from earth quickly enough to effect a change in their perceived colour? Jenkins and White (6) argue that the sort of speed at which a galaxy would have to be travelling in order to lower the frequency of the light coming from it to the red end of the spectrum is almost certainly impossible. By maintaining that the further away a galaxy is the faster it is moving, supporters of galactic red shift are suggesting, perhaps unwittingly, that distant galaxies are travelling at speeds in the order of 90 million mph - a suggestion that few would accept. If the galaxies were moving at anything like the sort of speeds that would be required to cause Red Shift the size of the Universe would have to be re-estimated to perhaps thousands of times larger than it is thought to be today. Also, if galaxies were travelling as fast as Red Shift theorists maintain, our night sky would not be such a crowded place. In order to change their apparent colour from yellow to red by movement alone galaxies would have to be travelling at speeds approaching 100 million mph. And, if Red Shift supporters are right in their assertion that the galaxies are continuously accelerating by now most of them would be invisible. At speeds of much more than 100 million mph the wavelength of the light coming from the galaxies would be stretched beyond the visible part of the electromagnetic spectrum and into the infra-red end of the visual spectrum. In other words, the light from the galaxies would be pulsing at rates that cannot be detected by the human eye and brain and, therefore, become invisible. So, if the red appearance of galaxies is not due to their moving rapidly away from us, why do they appear red?

The light that comes from a star can be impeded in its journey through space by gas clouds, magnetic fields, gravity fields and general debris. When light is slowed its pulses arrive at our eyes at a slower rate, effectively a reduction in frequency, and lower frequency light appears red. A good example of this is a red sunset. When the sun is low its rays have to pass through much more of the Earth's atmosphere than it does when it is high. This added resistance to its travel effectively reduces the frequency of the light and makes the sunset reddish.

The bluish look of some stars indicates the very high temperature at which these stars are burning. The light from bluish stars is little affected by travel because most of these stars are in our own galaxy and so this light has a relatively clear run on its way to Earth. So, if the colour of galaxies is not caused by movement is the Universe expanding? Even if the reddish appearance of galaxies has nothing to do with movement the galaxies could still be moving away but steadily and much more slowly than is generally maintained by most theorists. In other words, the Universe might still be expanding but not as rapidly as Red Shift theorists suggest.

Einstein's Special theory of relativity, published in 1905, suggests that space and time are one and the same thing. This also suggests that the Universe is expanding. And the later General theory added to this. Edwin Hubble's observations of the 1920s seemed to support Einstein. However, Einstein's maths and Hubble's real observations are in fact contradictory. An expanding Universe tells us that back in time the Universe was smaller than it is now. Reverse time more and more and the Universe gets smaller and smaller. This is consistent with Einstein's Special theory. When the Universe is, in theory, as small as it can get, that is, when it is a single particle, time can go back no further. This, together with Hubble's observations, has led cosmologists since the 1940s to believe that the Universe must have started as a result of a huge explosion. However, Hubble did not simply find that the Universe was expanding - his observations also suggested that it was expanding at an increasing rate, that is, that the further away a galaxy is, the faster it is receding. This is inconsistent with the General theory.

The General theory of relativity postulates, mathematically, that light is deflected by gravity. A major consequence of this is that the Universe is closed. In other words, neither light nor anything else can go beyond the Universe's outer edge. So, if a spacecraft takes off in any direction and travels far enough it will eventually come back to its starting point. This certainly coincides with the Special theory which suggests that the outer edge of the Universe is also the frontier of Time. In the Einsteinian scheme of things the future is out there, just waiting for the Universe to expand into it - as it were. Observations of galaxies in a closed universe would indicate that the distance between them was increasing and that effectively the Universe was expanding. However, the observations would also indicate that every galaxy was receding and getting further apart at the same speed, and this clashes with Hubble. The two models that cosmologists often use to explain the expansion of the Universe can also be used to describe Einstein's theory and Hubble's observations. The way the Universe is expanding according to Hubble can be represented by what happens during the baking of a raisin cake. Before baking the raisins are close together. During the bake, the raisins inside, which represent the galaxies, move out and away from the centre and from each other - those towards the outside of the cake moving further and faster than those nearer the centre. A rough guide to Einstein's Universe may be gained from the balloon model. Here, discs representing the galaxies are stuck to the surface of a balloon. As the balloon inflates the discs get further apart. And, of course, as the balloon gets bigger the galaxies get correspondingly further apart. This suggests that real observations of the night sky will show that the galaxies are receding and getting further apart. These two aspects of the Einstein model are in line with the Hubble findings. However, the Einstein model also suggests that the galaxies are receding at the same speed and moving apart at the same speed. In other words, Einstein's expanding Universe differs slightly from Hubble's. So, who is right? They are probably both wrong.

Hubble seems to have given today's scientists a means of measuring the age of the Universe, currently estimated at 16 billion years old. But is this figure accurate? By measuring the speed at which the most distant galaxies are receding and how far away they are researchers can calculate how long these galaxies have been travelling for. And, this automatically indicates the age of the Universe. However, if the Universe is expanding, an age calculation is always out of date.

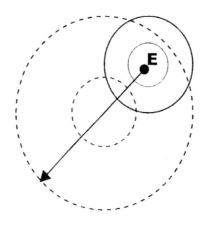

Figure 3

No scientist maintains that astronomers are able to see the edge of the Universe. However, it is claimed that two very large telescopes that are soon to be built will at last allow researchers to study the Universe's outer limit. But what does this mean? If the Universe is finite there has to be two edges. So, which edge do they mean? Figure 3 is a cross-section of how the Universe should look if it started with a bang. The outer dotted circle represents the outer boundary and the inner dotted circle indicates the edge of the void that should exist after 16 billion years of expansion. We can assume that our own galaxy is somewhere in the thick of things. This is indicated by the solid dot, labelled E. The smaller solid circle indicates the extent of present observations by astronomers. The larger solid circle suggests the distance to which the new breed of very large telescopes would reach. And these are the two distances that are used and will be used to calculate the age of the Universe. However, the arrow indicates the range that telescopes *would* have to have to get an accurate measure of the age of the Universe. But of course, figure 3 is a snapshot of the Universe. And, if as

Einstein's and Hubble's work suggests, the Universe is expanding, it is getting significantly bigger by the hour. If the Universe is 16 billion years old, and during this time the galaxies have been rapidly receding, the distant galaxies could be many billions of miles further away than they appear today.

The light from the most distant galaxies has taken and is taking several billion light years to reach Earth. These galaxies therefore look as and where they were that length of time ago and today they will be beyond the reach of the biggest telescopes imaginable. In other words, the distance to the outer edges of the Universe is unknowable. And, consequently, so too it's age. There is no way of knowing how much there is beyond what we can actually see. Also, if the Universe is expanding, there is no way of assessing how fast things beyond the range of telescopes are moving. Einstein's and Hubble's ideas on the nature of the cosmos are inevitably lacking in respect of size. We do not know the size of the Universe and probably never will. Does it have a size? Isaac Newton probably got it right when he concluded that the Universe must be infinite and must have always existed.

The notion of an expanding Universe raises two simple questions – expanding from where, and into what? If every galaxy is getting further away from every other galaxy, all must be travelling on radial paths that have a common point of origin. And, by simply projecting backwards, in respect of distance and time, this point of convergence should be readily identifiable. In the night sky of an expanding universe that is purportedly finite, and whose outer edge is reckoned to be not that far beyond the range of telescopes, the point that everything came from should be glaringly obvious. And, yet, nobody seems willing to even speculate as to an area that could possibly contain the point from which the Universe grew. Most cosmologists maintain that they can pick up the noise made by the Big Bang, and refer to this as background radiation. They state that these radiations come to Earth from every direction. This does not make sense. When asked as to how the noise of the Big Bang can be coming from all directions most theorists explain that the Big Bang was not a normal explosion. Also, even Big Bang supporters would not argue that the galaxies are creating their own space as they move away. And, if space exists at all it can only be endless.

In the context of a universe whose extent is unknowable it seems that research today largely involves cosmologists dreaming up and putting forward theories that they would like to be true.

Four

COLLISION COURSES

The first particle accelerators, built in the 1930s, produced valuable results especially in the field of medicine. However, the very large colliders that were built in the 1980s have since looked increasingly white and elephantine. Their story, when they are used to attempt to replicate the origins of the Universe, is one of experiment after experiment producing confusing and inconclusive results, and it seems that almost every experiment contradicts the results of just about every previous experiment.

Most often when scientific data are consistently confusing and contradictory, it is because the experimental designs or the experimenters' expectations are wrong.

Many of the experiments carried out in large accelerators today are associated with the theory known as the Big Bang, and are essentially attempts to recreate what is thought to have happened in the first few seconds after the Universe burst into existence. Most physicists believe that the Universe began with a huge outburst of energy and heat and that particles collided with other particles to create new particles. This creative process continued making, within just a few seconds, all the matter that exists in the Universe today. And particle accelerators are now often used to try to repeat this early creative activity.

A particle accelerator is basically a tube which can be quite small, for use in a laboratory, or very large. It can be circular, anything up to 27 kms in circumference, (built underground), or straight, but in both cases the principle is the same. A very high voltage is used to push millions of electrons or protons out of a suitable material. Very powerful electromagnets, placed at strategic points along the apparatus, are then used to accelerate these particles up to almost the speed of light. At the same time, particles are accelerated in the opposite direction, so that at a special point on the apparatus the two beams of particles collide. Here an observation chamber allows what is happening to be photographed by cameras with microscopic lenses. The snapshots that are taken in these 'how-the-Universe-started' experiments are, however, far from black and white. They require interpretation. And what the experimenters claim they are seeing in these experiments itself requires interpretation, and we could be forgiven for wondering whether the experimenters are merely stooges in an alternative experiment testing a hypothesis which has less to do with subatomic particles and more to do with naked kings and new suits of clothes.

The photographs taken in particle accelerator experiments have been likened to modern art paintings. They contain lots of straight lines, squiggles, spirals and dots. And, the claim made by particle physicists is that the lines are the traces that particles moving through the observation chamber make and that some of the dots are new particles created by the collision of two of the experimental particles - claims that seem harmless enough until we consider that the sub-atomic particles in question could be many thousands of times smaller than the nucleus of an atom, and that a microscope powerful enough to allow us to see an atom has yet to be built.

But could physicists really be so apparently blinded by the light of a big idea? Yes they could. In the early 1950s the design of particle accelerators was altered to counteract the slowing of time that is predicted in Einstein's Special theory of relativity - a slowing that as we saw in paper 1 is entirely a visual illusion. There are undoubtedly many good uses particle accelerators can be put to. But whatever it is their operators are seeing in the observation chambers of very large machines it is not particle colliding with particle.

Another major weakness in the design of collider experiments is that it is similar particles that they send round the apparatus towards the collision point. Normally, they propel protons or electrons. However, protons repel protons and electrons repel electrons. So, what happens in the collision chamber? As a particle approaches another particle the two will powerfully repel each other, making collision impossible. And, this further begs the question of what is actually happening in collider experiments.

Five

THE BIG DAMP SQUIB

The theory about the creation of the Universe commonly referred to as the Big Bang has mathematical, observational and experimental support. And, perhaps simply because it has a catchy label, it is much loved by the media. However, despite its popularity the Big Bang theory is not without its problems.

The Big Bang theory does not readily fit with research work that has been, and is still being, done into establishing the age of the Universe. It brings with it its means of measuring the Universe's age and the figure this produces does not always agree with the figure calculated by cosmologists studying the age of the stars - not always, because the Big Bang figure can be varied.

If the Universe is a finite size, albeit an expanding one, and if it started with a huge explosion, the time it has taken to reach its present extent tells us exactly how old it is. The trouble is, the rate at which the Universe has expanded is difficult to ascertain. If it expanded very rapidly the Universe is 'young'. If it expanded comparatively slowly the Universe is 'old'.

The key figure in Big Bang calculations of the age of the Universe was provided by Edwin Hubble. This is referred to as a constant and links the speed at which a galaxy is receding from Earth with its distance from Earth. Today, it seems that most cosmologists disagree as to how big or small a number the Hubble Constant should be. Mostly, researchers disagree about how far away the most distant galaxies are, but this debate is a futile one. There is no way of knowing the distances to the furthest galaxies. If Hubble's working assumption as to why some galaxies appear reddish was correct, he was unwittingly studying and calculating distances to galaxies that were long gone and well beyond sight. He was merely seeing the light from these galaxies – light that had taken billions of years to get to him. And, if the Universe is expanding, the actual galaxies in question, would have been, as they still would be today, many billions of miles further than his calculations suggested. There are an unknowable number of galaxies out there, on the move or not, whose distances from Earth are unknowable. Consequently, the age of the Universe is unknowable.

The inspiration for the Big Bang theory came largely from the results of Edwin Hubble's research. However as we also saw in Paper 3, Hubble's findings are inconclusive and so provide only rather shaky support for the Big Bang.

The mathematical support the Big Bang has comes from Einsteinian relativity which states that space or distance and Time are practically one and the same thing. According to Einstein the Universe has expanded with the progress of Time, and this means that the outer limit of the Universe is also the frontier of Time. If then we mathematically or theoretically reverse Time, space and distance shrink accordingly. If we continue this process the Universe gets smaller and smaller until it becomes a single speck - a point which also marks the beginning of Time. The distortions in Time that Einstein predicted and attributed to distance and space travel, however, are, as we saw in Paper 1, entirely illusory and cannot be used to support a theory about real events.

And, of course, simply stating that the Universe began with a bang is not enough. How matter, now in the shape of billions of stars, came into existence, has to be explained. This was duly attempted and by 1948 the first version of the Big Bang theory had been put together. Projecting aspects of Einstein's Special theory backwards produced the starting point - a situation of no space and no Time. What is reckoned to have happened immediately after the

Big Bang occurred was inspired by contemporary research projects in which atoms were being split in small particle accelerators. Matter, it is postulated, was created with and by the explosion that occurred through particle colliding with particle. And right away researchers set about trying to replicate what happened in the first few seconds of Time using particle accelerators.

The results of particle accelerator experiments designed to test the Big Bang theory have never been satisfactory. In the early days this was put down to a lack of power and so bigger and more powerful accelerators were built. But the confusing and conflicting results continued.

We do not know if there ever was a state of nothingness. The postulation that there was, as indicated in Einstein's Special theory, is based on a misconceptualization of Time and a faulty fusion of Space and Time into a continuum.

The loss of its marking the beginning of Time greatly weakens the Big Bang theory but it does not rule out the possibility that at some point in time there was a huge explosion. The big claim of the Big Bang, though, is not that there was a big bang, it is that an explosion was the means by which all matter was created. The Big Bang, therefore, has to be no ordinary explosion. Having built and tested many atomic bombs we know quite a bit about normal explosions. Also, everywhere in the Universe stars are exploding as part of their normal pattern of life. The essential nature of a normal explosion is that everything within its sphere of influence is instantly pushed out and away in every direction from its centre. No creation takes place in this kind of explosion. So, what was it about the Big Bang explosion that made it behave in a way that is inconsistent with today's laws of physics? The Modern Physics answer to this is that, at the beginning of Time there were levels of heat, energy and pressure, which are impossible today. But there cannot be any levels of anything without atoms. To overcome this problem, more recent versions of the Big Bang theory postulate an explosion of several stages.

The final stage of the Big Bang is the point at which its immense pressure is instantly released and its heat dissipated, throwing out at the same time sufficient matter to eventually build the entire Universe. In the first few stages, which together span just a few seconds, there was an outburst of immense heat and energy, the creation of particles, and a huge build-up of pressure. This explanation of how the Universe was created certainly sounds more scientific than the classical perspective which is that the Universe has simply always existed. But this ostensibly more scientific explanation of how everything came about cannot explain what caused the initial spark. It can only claim, rather unscientifically, that about 16 billion years ago there was a huge explosion which created everything from nothing.

There is no real evidence that the Universe came into existence as a result of a massive explosion, and the only real support for any assertion that it did comes from the Hubble data that suggests that the Universe *might* be expanding.

Perhaps the appeal of the Big Bang for researchers is its suggesting that there has been one occasion in Physics when we got something for nothing.

* * *

The Big Bang is undoubtedly a very popular and widely accepted theory. However, it has many weaknesses. Most scientists believe that about 16 billion years ago the agitation that had been building up in a single particle of matter reached a critical point. The heat and pressure reached a level that was no longer containable and the particle exploded. This is what has come to be known as the Big Bang. But it was no ordinary explosion.

39

According to theorists, the Big Bang started everything. Before it there were no stars or planets, no Life, no Space, no Time. The instant the explosion occurred marked the start of Time and as the fragments and the blast wave moved out and away from their point of origin they created Space. The early expansion was extremely rapid, and within just milliseconds the fragments of matter that came from the original bang began to collide. In doing so they created more matter. The expansion and the collisions continued until within minutes the matter of the entire Universe had been created.

At some early point this huge ball of activity ignited and the Universe became a gigantic fireball. The expansion continued but began to slow down, and the enormous temperature that had existed began to drop. Very gradually, over perhaps a million or so years, the stars began to form and as they did so they clustered into galaxies. Eventually, the Universe, more or less as we know it today, had taken shape.

There are several versions of what is known as the Big Bang theory. Since it was dreamt up and aired in the 1940s parts of the theory have been amended, new ideas have been added and bits have been completely removed as successive scientists have tried to improve a construction that was always shaky. When this happens it is almost invariably the case that the theory in question has been based on a false premise. The Big Bang is just such a theory - it is fundamentally flawed. And no amount of crack-filling and hole-plugging can make correct a theory that is basically wrong.

The biggest question about the Big Bang has always been the one that demands a satisfactory explanation as to what existed before it happened. And the reply is always the same. Prior to the Big Bang there was absolutely nothing - no matter, no Space, no Time. But this is inconsistent with the fact that it was a particle of matter that exploded. So, something *did* exist before the Big Bang. And, how did this small piece of matter come to be? Nobody really knows the answer to this, and few theorists even speculate. Also, how long had the particle in question been in existence before it exploded? The fact that it existed at all means that the Big Bang was not the start of Time. At least one theorist tries to eliminate this contradiction by suggesting that the particle existed for less than a billionth of a second. But this is naïve. Science is as concerned with billionths of a second as it is with billions of years. Striking a match in a gas-filled room with the intention of just having it lit for a short spell is not recommended. In physics, very small amounts can be very significant. Also, if it all began with a particle this had to be taking up space - albeit a very small amount. So, the Big Bang was also not the creator of Space.

There is also the problem of how a single particle of matter gets itself to an explosive state. As we know today, the high levels of pressure and temperature necessary for an explosion are reached as a result of nuclear or chemical reactions within and/or between certain elements. This begs the question of what the original piece of matter was made of. To cause a nuclear explosion the piece of matter would have to contain two appropriate elements such as hydrogen and helium. So, again, from nowhere into existence appear two elements that just happen to be suitable for a nuclear explosion.

A normal explosion occurs when the heat and pressure within a confined space reaches a level beyond which these can no longer be contained. This build-up is caused by chemical reactions. In a single particle there is no opportunity for any sort of reactions to take place. In any case, the explosion of a subatomic particle is logically impossible. When something explodes it fragments. If a particle were to explode it would have to shatter into many small pieces. And, if there were just a hundred of such fragments this means that there would exist pieces of matter one hundred times smaller than the particle that was, prior to the explosion, the smallest possible piece of matter.

Recent versions of the Big Bang theory try to overcome this problem by postulating that it was in fact a piece of matter the size of a pea that exploded. The introduction of the first piece of matter as pea-sized certainly gives the theory something that is capable of exploding but it by no means eliminates all the problems associated with the earlier single particle version of the theory. For a start, we now have a sizeable chunk of material popping into existence from nowhere rather than the tiniest possible particle.

By introducing the proposition that the Universe's starter was a small piece of matter, Big Bang theorists have introduced the explosive potential that was lacking in earlier versions of the theory. A small piece of matter comprises many millions of atoms, each of which contains a huge amount of potential energy. But, where did they come from? And where did the nuclear force that gives the atom its structure come from? Also, still without having extracted as much as a speculative guess from theorists are the questions of why the temperature and pressure within a small piece of matter would spontaneously begin to rise and keep on rising, and how enough energy was created from a pea-sized bomb to fuel the fires on countless billions of stars.

Energy cannot be created or destroyed. Energy is stored in many ways in nature and is used to do countless billions of jobs every day throughout the world. When used, energy is difficult to recover, but it is not lost. In other words, the amount of energy in the Universe is constant and has been so for however long the Universe has existed. So, any theory that proposes that the Universe had a tiny beginning has also to explain the creation of a universeful of energy. Some supporters of the Big Bang would argue that the piece of matter that exploded was infinitely dense and that the infinite energy associated with this was dissipated along with the expansion of the Universe, but this is fundamentally flawed. Infinite density is not physically possible. It is a purely mathematical concept. Matter cannot be destroyed or compressed. Every given piece of material has a maximum density. Some theorists guess that the energy of the Universe would have been created at the same time as, and as a result of, the creation of the mass of the Universe. But this too is problematic.

When an explosion takes place the resulting fragments move radially out and away from the source of the explosion getting further apart as they do so. This can also be understood by imagining a confetti-covered balloon being inflated. When the balloon is the size of a tennis ball the pieces of confetti are tight together. At the size of a football the small pieces of confetti are now more than an inch apart. If the balloon reaches the size of a beach ball the pieces of confetti will be several inches apart. The bigger the balloon gets the further apart the pieces of confetti get. This is what would have happened to the fragments of the Big Bang, and it is a problem for a theory which claims that the creation of matter was the result of a new particle being created when two particles collided. And, that this process continued, it is argued, until the matter of the entire Universe had been created. But, how were particles travelling along a radial path able to get on to a collision course in the first place? Fragments blasted away from the source of the Big Bang would have been miles apart in no time. Also, during this matter-creating phase of the Big Bang, every particle would have to have collided many billions of times over with other particles. This is an unexplainable creative capacity - a universe from nothing. Also, it is reasonable to doubt whether the momentum that was given to particles by the Big Bang would have been sufficient to keep collisions going long enough to create the matter of the entire Universe. Every collision would have dissipated energy and taken some of the steam out of the speeding particles.

It has to be assumed, since it is not stated in the theory, that the particles that were involved in the collision phase did not have a charge, that is, there were no positively charged protons or negatively charged electrons. Had the particles that were flying about at this stage been charged the collisions of the sort that were required would have been impossible. Like charges would have repelled each other and unlike charges would have, quite simply, stuck together. And yet, charged particles were required at this stage to keep the Big Bang going. Electrons, protons and neutrons structure the atom, and it would have been a variety of atoms, that is, atoms that differed according to the number of particles they comprised, that would have, as a result of their interactions, completed the construction of the Universe. But how could atoms have been formed? Atoms require charged particles. How did particles get their charge? Also, if current thinking on the structure of the atom is correct, a strong and a weak nuclear force is also required to hold the atom together. Where did the two nuclear forces come from?

According to most Big Bang theorists, there was a period during which the early Universe was a fireball. But before such a phase could have begun certain things would have to have been in place. The spontaneous start of fire would have required the presence of suitable elements such as hydrogen and helium. Where did these come from? Atoms cannot be simply thrown together. They are precisely structured and finely balanced systems. The shape and stability of an atom is maintained by a delicate balance of its internal forces. This could not have been achieved by chance or in a piecemeal way. For example, how was a cluster of protons, which form the nucleus of the atom, able to come together and stay together? Like charges repel. Right away, here, the strong nuclear force is required. Also, there is a powerful attraction between the positively charged protons and the negatively charged electrons of the atom. So, why do the electrons not collapse into the nucleus? Furthermore, how are the electrons which powerfully repel each other able to form, in close proximity to each other, the outer shell of the atom? What all this is saying is that the atom could not have been put together by accident. Nobody knows how atoms were formed.

Today, we know that stars burn as a result of a chemical reaction normally between hydrogen and helium, and that this reaction, most theorists believe, is sparked off by the strong pull of gravity within the star squeezing the atoms of these elements tightly together. And, presumably the fiery phase of the Big Bang would have been set off in a similar way. That is, the huge cloud of gas that existed would have been drawn very tightly together by gravity. Where did gravity come from? The force of gravity is still a mystery today.

After the fireball, most versions of the Big Bang state that there was a period of rapid expansion. But what followed this is anybody's guess. Presumably, pockets of gravity drew in and captured a share of the gasses that were around and thus created the first stars. Later, somehow, stars themselves were drawn together into the galaxies that we see today. And, there you go - a universe.

The above discussion can be boiled down to a one-word description of the Big Bang thesis - nonsense. It flouts the laws and principles of physics, insults common sense and seems more the product of wishful theorising than the result of evidence-based reasoning. Why then, if it is such a bad theory, is it so popular? The reason for this starts in the late 19th Century with the discovery that light travels at a measurable speed.

Albert Einstein and several of his peers realised that, where very long distances are involved, the slight delay in light getting from A to B could have weird effects. They calculated that observers of very distant objects and events would see things that could not be explained by the current laws of physics.

In 1905, Einstein published his famous paper on Special relativity. In this, Einstein's calculations indicated that science's current concept of Time would have to be demolished. If Einstein's calculations were correct Time would now have to be regarded as a variable, and space and Time would have to be understood as one and the same thing. And if Time is steadily moving forward distance too has to be advancing. If distance is generally increasing the Universe must be expanding. This was powerfully supported by observations made by the astronomer Edwin Hubble in 1929. Hubble reported that every one of the galaxies he had studied, in every direction, was receding. If the Universe is currently expanding, at some point earlier in time it must have been much smaller than it is at present. And a smaller and smaller Universe means that at some point it would be the smallest it could possibly be - a single particle. And with the concept of space-time in mind this point would also be the start of Time. Before this there would have been no Time and no Space. Then, to get an expanding Universe, it is reasoned that this single particle must have exploded, and it is calculated that this would have taken place about 16 billion years ago. On hearing this in a discussion in the early 1950s the scientist Fred Hoyle coined the phrase 'Big Bang' and this has stuck, perhaps due to its popularity with the mass media. Undoubtedly, the scientists of the day, found that the sequence of events that is the Big Bang, although revolutionary, entirely credible and acceptable. It was neat. Everything seemed to fit. And today, almost all scientists fully support the notion of the Universe having started as a result of a huge explosion.

Part of the popularity of the Big Bang is largely due to the high esteem in which Einstein is held. It is also due to the full acceptance of Hubble's interpretation of his observations. The trouble is, as we have seen, Einstein was wrong and Hubble may not have been seeing what he thought he was seeing.

Six

SPOOKY ACTIONS

Quantum physics is all about the atom and it's particles, and the goal of quantum researchers is to understand and utilise the apparent weird behaviour of sub-atomic particles in certain circumstances. There was a significant rise in the level of interest and activity in all things quantum during the first two decades of the 20th Century and, undoubtedly emboldened to some extent by Einstein's postulations on the nature of the relationship between space and time, to explain some of the quantum phenomena that simply do not make sense, researchers have come up with theories that necessarily require the abandonment of common sense and the rejection of some of the central tenets of classical physics. Despite this, quantum physics is very popular and attracts great interest throughout science. But, as is always the case, we abandon common sense at our peril.

Early landmark quantum experiments seemed to demonstrate that two sub-atomic particles could become bonded or entangled. It was found that whatever was done to one particle a second particle mimicked exactly the way in which the first particle responded. It was reasoned by the experimenters that, if and when two particles became bonded, the distance between them should be irrelevant. In other words, if a bond existed between two particles one metre apart this bond should still function when the particles are a kilometre apart, and beyond. However, they fairly soon realised that they had overlooked the significance of the relationship between distance and time.

Several experiments have been carried by experimenters based in Switzerland designed to test Entanglement theory. Generally, photons are shot from a laboratory along a fibre-optic cable to a monitoring station 25km or so away. Typical of quantum experiments, the results of these tests have been frustratingly inconclusive. A problem that has been highlighted during this series of experiments is the tiny delay in the arrival of particles at the receiving station. The theory of bonding requires that whatever is done to one particle is instantaneously done to the other. In the Swiss experiments this is not happening. How long will it be, we may wonder, before our dauntless researchers begin to suspect that their experiments might be based on a false premise?

As well as Entanglement, where two particles, even at a great distance apart, behave as one, Quantum theorists also believe that one particle can be in two places at the one time.

An early study by Thomas Young in 1801 was designed to establish and demonstrate that light travels as waves. Here, two screens were erected on a bench, facing in the direction of the sun, one about a metre behind the other. In the screen nearest the sunlight two narrow vertical slits were cut to allow the sun's rays to pass through and land on the rear screen. The pattern that appeared on the rear screen comprised vertical bars of light, and Young concluded that this was evidence of light travelling as waves.

Roger Penrose describes a more recent and slightly more sophisticated experiment designed to investigate the wave nature of light (1). Here, the apparatus comprised two vertical screens one metre apart, an electric lamp which produced monochromatic light (yellow), and a photon counting device. Penrose mentions the fact that this type of bulb radiates 100,000,000,000,000,000,000 photons per second. In the screen nearest the lamp two very narrow slits were cut, very close together.

When one slit was blocked in the above set-up so that the light passed through just one slit, it fanned out by the phenomenon known as diffraction. When the light was at full strength the pattern on the rear screen was a uniform pattern with some evidence of some photons having been deflected by the edges of the slit. When the strength of the light was greatly reduced the light on the rear screen became apparently made up of individual spots. Penrose states that this pattern is created by the arrival at the rear screen of streams of single photons. When the second slit was opened, and the light was restored to normal full strength, things changed significantly. A bar pattern appeared on the rear screen, that is, there were alternating bands of strong and slightly weaker light. This is known as an interference pattern, which Penrose describes as having a waviness to it. The overall brightness of the light on the screen is now twice what it was before, at some points it is four times as strong as it was before. Also, where light was faint in the one slit situation these areas are now dark. Penrose suggests that when just one slit is open photons were allowed to land in these areas but that when both slits are open something was preventing them getting to these areas. Penrose asks, " How can it be that by offering the photons an alternative route we have stopped them taking either route?" And, " How can a photon know, that when it passes through one of the slits, whether or not the other slit is open?" He states that the slits can be any distance apart for the cancelling or enhancing phenomenon to take place. Penrose states that the photons seem now to be behaving like a wave, rather than as particles. He argues that the cancelling and enhancing that takes place here is a familiar property of ordinary waves. He explains this by stating that if a wave passes through each slit these will add on the other side if they are in phase, that is, when peak matches with peak and trough matches with trough. On the other hand, if a peak matches with a trough these cancel each other out, and produce no light on the screen. This is described as wave disturbance. A disturbance, Penrose points out, can pass through one slit or the other. Here though, he explains that things are different. Each individual photon behaves like a wave entirely on its own, and passes through both slits at the same time. This means that it can interfere with itself. The experimenter knows this because by adjusting the strength of the light to very low he ensures that only one photon at a time is arriving at the slits. So, when only one slit is open the photon passes happily through that slit. When the other slit is open the photon again happily passes through that slit. However, when both slits are open one photon can pass through both slits and cancel itself out on the other side of the screen, thus producing a dark patch on the rear screen. So, it is not that a photon sometimes behaves as a wave and sometimes as a particle it is simply that a particle can behave as a wave entirely on its own. Where there are alternative routes sometimes photons add and sometimes they cancel each other out.

Penrose states that in the 'double-slit' experiment it is not the case that a photon splits so that half can pass through one slit and half through the other. He argues that the use of a particle detector proves that a complete photon passes through either one slit or the other. However, when the detector is placed at one slit so that the experimenter knows for certain which slit the photon has passed through, the wavy interference pattern at the screen disappears. So, for interference to take place it seems there has to be a lack of knowledge as to which slit the photon actually passes through. To get interference, therefore, a photon has to pass through both slits, sometimes adding and thus reinforcing itself and sometimes subtracting and so cancelling itself out. This is an extraordinary claim. Can it be true?

The correctness of the conclusions that quantum theorists have come to from the findings of double-slit experiments rests in turn on the accuracy of the assumptions that are made about what is actually going on in these experiments. How well does the experiment described by Penrose stand up to this sort of scrutiny?

We know the speed of light and that it pulses as it travels, that is, at any point in its journey its strength fluctuates in a regular way. So, each pulse of light involves a steady rise in strength to a peak or a maximum and a gradual fall to a minimum, and the number of pulses per second or over a given distance is referred to as the frequency of light. Even when passed through a prism or a volume of water the speed of light varies very little, and unless it is deliberately altered the frequency of light is also very stable. Within the relatively narrow visible spectrum of light it is the frequency that gives rise, within the brain, to the creation of colour. The strength of light varies to some extent with the number of photons involved and, here again, this is very stable. When we are talking about photons we are talking about billions. Since our seeing anything requires a sufficiently strong stimulus at our retinas even a small drop in the strength of photons arriving at our eyes can mean that the light literally goes out. Reading a newspaper in very dim light, for example, requires the reflection, from page to retina, of several billion photons from every letter. Some idea as to the size of a single photon may be gained from considering the fact that several thousand billion photons could be placed on an area equal to the cross-sectional area of a human hair.

Above, Penrose describes a point in the experiment where the light is turned well down and there is only one slit open. He describes the pattern on the rear screen as appearing to be made up of dots, and suggests that this indicates the arrival at the rear screen of single photons. This demonstrates Penrose's apparent poor appreciation of the nature of light. In order to simply perceive the pattern of light in question as a collection of dots means that every dot must be reflecting billions of photons. The broken nature of the pattern is being created by the very weak light being scattered by the edges of the slit. Remove the front screen and the pattern of light on the rear screen would return to a uniform patch.

The convention of thinking about and referring to light as waves has come about because of the way in which the strength of light fluctuates as it travels. In the space of just one metre, light can rise to a maximum and drop to a minimum anything between one and a half and two and a half million times. We think of the strength of light at these points as rising and falling, and when these values along a ray of light are plotted onto a graph they produce a sine wave. Does this mean that photons travel as if on a roller coaster? No, not at all. In reality, photons do not physically go up and down, they travel in straight lines. In mathematics light is invariably represented on a graph as a wave or by a wave producing formula. A peak in a graph representing the strength of light is caused by a pulse of tightly packed photons. At a trough, the photons are much more strung out. Many scientists, however, believe that, as they travel, photons rise and fall physically.

In the double-slit experiment, when the light is at full strength and both slits are open, Penrose describes the pattern of light on the rear screen as having a waviness about it, and states that this is being caused by the mutual interference of the two sets of waves of light that have come through the two slits. The theory here is that when the waves are in phase, that is, when peak is matching with peak and trough with trough, the light combines to produce very bright light. And, when the waves are out of phase, that is, when a peak is matching with a trough, the two cancel each other out and so produce a band of darkness. There is no doubt that in the double-slit experiment very bright, less bright and dark bands are produced on the rear screen, and the experimenters have no doubt that this is being created by particles behaving as waves. But, if particles go physically up and down as they travel surely something must be making them do so.

In nature, waves occur on the surface of liquids when a force is applied. The example that scientists most often give to describe how light travels as waves is the dropping of a pebble into a pond. Here, the force of the pebble entering the water is met by the resistance put up by the weight of the water, and this results in concentric circles of waves being pushed outwards. The energy of the waves is finally dissipated at the edge of the pond. But, the characteristic way in which light pulses as it travels is not the result of the progress of the streams of photons that comprise light being resisted and thus being squeezed into physical waves like the waves on the surface of an expanse of water. Light pulses as it travels through a vacuum, where there is no resistance. And, its photons travel in straight lines. They have to. There is no way rolling streams of photons could create the extremely sharp retinal images that the brain needs in its providing us with a clear, detailed and accurate representation of our surroundings.

As we have seen, passing a fairly strong beam of light through two narrow slits creates a pattern that is evocative of waves. It seems that it is simply assumed that this evidences the wave nature of light. But is this assumption correct? Is there a possible alternative explanation for the pattern that appears on the rear screen? In the double-slit experiment, simply by passing a strong beam of light through two slits we could reasonably expect the production of two solid bands of light on the rear screen. However, on emerging from an opening that is small enough to deny its free and full passage a beam of light inevitably diffracts. An important aspect of this is that the two main beams coming from the slits partially merge. So, as an alternative to the wave explanation, it can be reasonably argued that it is this merger that causes the patch of very bright light on the rear screen. The areas of the beams that do not actually overlap produce the patches of light of normal brightness either side of the very bright patch. The smaller fuzzy patches on the outside of this central patch can be explained as light deflected from the edges of the two slits with the dark patches in-between simply being due to the fact that the deflected light is also displaced slightly. The entire pattern on the rear screen, therefore, can be explained in terms of pulses of photons travelling in straight lines. Indeed, any situation where light appears to be travelling as waves can be explained in terms of pulses of photons travelling in straight lines.

The diagram at Figure 4 shows how photons travelling in a straight line can produce a wave pattern when the energy at a particular point in the stream of photons is measured and plotted. The diagram, of course, is not to scale. For example, the distance between pulses, indicated by the cluster of dots, is measured in nanometres. Also, even in a very narrow beam of light there will be billions of streams of photons tightly packed side by side. In other words, at the highest energy points in a stream of photons, represented in the diagram by three dots close together, could be representing countless billions of photons in a beam of light.

A more sophisticated version of the double-slit experiment involves the use of an interferometer. This is a device that can demonstrate the ability of one particle to travel along two separate paths at the same time. After having made their separate journeys the two particles are brought together again. At this point, experimenters claim, the particles create an interference pattern. The procedure involves sending photons through optical beam-splitters, which are essentially half-silvered mirrors. When a photon arrives at the beam-splitter it has a 50 percent chance of passing straight through and continuing along its original line of travel, and along a lower arm, and a 50 percent chance of being deflected onto

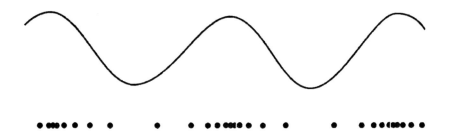

Figure 4

a route at right angles to its arrival route. Then, deflected by an angled mirror, sent along an upper arm. So, one particle will travel along a lower arm and a deflected particle will, via a mirror, pass along an upper arm. Experimenters argue that this is similar to the double-slit experiment in that during the experiment it is actually one photon that travels down both the lower and upper arm simultaneously. After travelling a short distance separately mirrors bring the 'two' photons together again. Turning the intensity of the light beam way down, it is argued, ensures that only one photon at a time is passing through the interferometer. Each photon thus finds itself in what is known as a superposition of travelling along two paths at the same time. It is not the case, the experimenters argue, that it is one particle that is travelling along the upper arm and a different particle passing along the lower arm. The ultimate creation of an interference pattern, it is claimed, is proof of one photon having interfered with itself. There is next to no point in describing this experiment any further. It is fatally flawed, in two main ways. First, a filament-type lamp is not capable of emitting one photon at a time, and there is no convincing evidence to suggest that any device can produce photons in isolation. And, second, in any case, just two streams of single photons could not create any kind of pattern, let alone an interference one. Our seeing just a small spot of light on a surface involves the arrival of many billions of photons at this small area. This unimaginable number of photons ensures that reflections from the spot go in every direction, including towards our eyes in sufficient quantities to create a good image on our retinas. Seeing a pin-head-sized dot necessarily involves the reflection and reception of billions of photons.

In the experiment immediately above, the design of the interferometer and of the experiment reveal a complete lack of understanding, on the part of the designers, of photons and the visual perception they give rise to. If the source of light is on at all and a pattern of light is being created on the screen there will be billions of photons streaming through both arms of the apparatus.

Normally, when the introduction of a new piece of apparatus seems to bring about a change in the outcome of an experiment the experimenter immediately suspects that it is the new bit of equipment that is responsible for the change. Not so in tests of Quantum theory. Here, when a detector is added, for example, experimenters tend to assume that it is the new knowledge that they have gained that has affected what happened during the experiment. They also believe that the particles that are involved in the experiment can gain knowledge of what is going on and that this can alter the outcome of the experiment. In the double-slit experiment, for example, researchers can ask questions about how a photon can 'know' that the other slit is open, or closed, or that they can 'know' that there is a detector on the other side of the slit. In other words, in quantum research, when an experiment produces strange or conflicting results, rather than suspect the experimental design or their pre-experimental assumptions, experimenters seem to be happy if not keen to go for a magic explanation.

In a recent repeat of the double-slit experiment a laser, rather than a spotlight, was used as the source of light and two small holes were cut in the first screen rather than two slits. Quantum theorists believe that electromagnetic energy comprises waves and particles and that setting up an experiment to measure one of these eliminates the possibility of measuring the other. In other words, in tests, an experimenter can measure particles or waves but not both. In the double-hole experiment the light from the laser diffracted after passing through the two holes and produced what the experimenter described as a normal interference pattern on the rear screen. Here, typically, there is a patch of strong light in the middle of the pattern with weaker patches of light on either side. Beyond this, on both sides there is a dark patch and at the two extremes there is a fuzzy patch of light. In agreement with double-slit experiments, the experimenter stated that the interference pattern was being produced because the light between the two screens was behaving as waves. That is, where peak was matching with peak the two waves were adding together to create the patch of strong light. On the other hand, where trough matched with a peak these two cancelled each other out to produce a dark patch on the rear screen. Adding and subtracting waves like this is a standard and simple procedure in mathematics, and quantum theorists seem to have no doubt that this can be used to explain an interference pattern. They are, however, entirely wrong. Certainly, light beams can be added together, but there is no such thing as negative or minus light. Light is either present or it is not, switched on or switched off. And the assertion that an area of light can be cancelled out by an area of no light is preposterous. Also the extent of the interference pattern that is typically produced in double-slit experiments is such that, if light waves were the cause, they would have an amplitude of several inches. Is it simply the case then that we cannot see this wave pattern because of the huge velocity of light?

The New Scientist of 9 April 2005 contains a report on an experiment in which it is claimed a special detector, with an aperture that opens and closes every billionth of a billionth of a second, captured light travelling as waves. The 'photograph' is certainly of something that looks like a wavy beam of light but the text reveals that it is not a proper photograph of light behaving freely. The experiment in question involves sending a beam of light into a chamber filled with neon gas. The detector is positioned such that it can fire single pulses of x-rays which image the waves. To be sure of which part of the wave is being imaged these pulses are synchronised with the light beam. The x-rays knock electrons from the atoms in the neon gas and these then move towards the detector. The magnetic field of the passing light beam either slows or speeds up the movement of these electrons depending on the strength and direction of this field. Thus, the arrival of each electron at the detector provides an instant 'picture' of the light beam on its journey through the chamber. These snap-shots, 6 million individual exposures of identical passing light waves, are added together to provide a whole image of the light beam.

Photographic terms are used to describe the key events in the above experiment, but there is really nothing in the procedure that is genuinely photographic. It seems that a fluctuating stream of electrons enter a detector creating a signal which is sent to an oscilloscope. As in any oscilloscope, a pulsing signal is represented on a screen as a sine wave. The image that is reproduced with the report suggests a wave amplitude of about 2 or 3 centimetres. This, and the wavelength that is suggested in the picture, are entirely inconsistent with a wavelength of light that is measured in nanometres.

A join-the-dot map was never going to be acceptable as evidence in the photon-particle debate on the nature of light. And, the scientific community has responded accordingly to the research project mentioned immediately above. A pencil-thin beam of light comprises billions of streams of photons packed tightly together. So, an isolated, substantial wavy line of light produced by the apparatus described above has to be a fudge. It seems that the beam of light that was passed into the chamber was bent or moved by the other forces that were applied to the apparatus. Presumably, then, electrons were bounced off the beam at many different locations providing perhaps millions of dots that could then be strung together in a wave shape by a computer. A genuine, direct photograph of light would be a very powerful piece of evidence in the debate about the nature of light, but the magnification that would be required here is probably unachievable.

When a beam of light is pointed towards a pair of pinholes or narrow slits and is such that it cannot pass freely through these openings some turbulence inevitably takes place at the edges. The light that passes relatively freely through the central part of small openings creates two fairly strong beams which diffract and, after a short distance, partially merge. On striking a screen these create a pattern which is bright at the centre and less bright on either side. From the edges of the openings, two, or possibly four, weaker beams of deflected light emerge. These will land well to the left and right of the central bright area leaving a dark patch in between. When one of the openings is closed and the strength of the source of light is greatly reduced the deflected beams can fade out of sight leaving only a diffracted main beam. A perfect and typical interference pattern can easily be created by beams comprising straight-line pulses of photons. The two pin-hole experiment above merely confirms that streams of photons can be made to produce a specific pattern of light and can be reflected towards a desired target.

If true, the ability of a particle to be in two different places at the same time is extraordinary, if not magical. It certainly defies common sense. But is it true? Quantum theorists not only argue that it is but also that it is not the only trick that a particle has in its repertoire. Any particle, they argue, is capable of forming a bond with another similar particle such that the two behave as one - no matter how far apart they are. This is referred to as Entanglement.

A novel way of testing Entanglement theory involved firing entangled photons across the River Danube (New Scientist, 28 June 2003). This was the first time that Entanglement had been tested for over long distances without the use of optical-fibre cables. The experimenters built a miniaturised robust version of the entangled photon generators that are routinely used in laboratories. This produced entangled pairs of photons in terms of the quantum property of polarisation. How did they do this? The researchers do not explain. Can a photon be polarised? Each one of the supposedly entangled photons of a pair was beamed to a receptor - one 500m away across the river, and the other 150m along the same bank that the transmitter was situated on. Telescopes were used on all devices. Transmitting telescopes were focused onto the receiving telescopes. Later, the receptors then focused the photons onto the photon detectors. The polarisation of each photon entering each detector was measured, and it was found that each pair was still entangled.

It is surely stretching credibility somewhat to claim that, in the 'Entangled Photons Dance Across the Danube' experiment, photons could be fired with pinpoint accuracy to receptors 150m and 500m away. Also, it seems that the research team decided that carrying out the experiment at night provided an adequate control against entry into the receptors of extraneous photons. But keeping out light out of doors is very difficult, even at night. Light from buildings, street lamps, the lamps that the experimenters may have been using at the transmission site and even any stars that may have been out could well have got into either or both of the receptors. The big question, though, as ever in tests of Entanglement, concerns whether experimenters really can isolate and control individual photons.

The concept of Entanglement was put together entirely by mathematical calculation. Its arrival in the 1920s generated a huge amount of interest among the scientific community, but there were many sceptics. Einstein was not at all impressed and even after observing an apparently successful demonstration of the existence of Entanglement argued that there had to be hidden local variables at work, and that the discovery of these would provide a rational explanation as to why particles sometimes appear to be bonded. Einstein described apparent Entanglement as, "spooky action at a distance" (3).

Early tests of Entanglement theory were simple and merely involved zapping the near photon of a purportedly entangled pair with a beam of light, looking to see what had happened to this one, and checking the other one, that is, distant one, to see if it had been similarly affected. The distant photon was normally 1 or 2m away. How the experimenters managed to isolate and hold a single photon in such a way that it could be struck by a beam of light, and how they managed to isolate and hold the distant photon in such a way that it could be affected by what happened to the near one, is never fully and clearly stated. It is claimed that many such trials were carried out successfully in the 1920s and 30s, but rather than look for a rational explanation for their weird findings in these experiments, experimenters invariably came up with answers that can only be described as magical. The possibility that 'stray' photons could be striking the distant photon occurred to some researchers and they controlled for this by reducing the strength of their source of light. This greatly reduced the number of possible false positives that the researchers had been recording, and the new figures that were now being recorded, it was argued, were because there were now fewer photons in the system. Certainly, reducing the strength of the light source would reduce the number of photons in the system. But, to what extent? Our being able to see a light beam at all tells us that billions of photons are being reflected from it, towards our eyes and in every direction. The research team subsequently reported that 97 percent of their positive trials were due to teleportation. How did the experimenters know the difference between a 'hit' that was due to teleportation and one that was the result of being struck by a stream of electrons? The assumption that was made here about the nature and behaviour of light, however, was false. Experimenters mistakenly believed that they could create a situation in which there were very few photons in the system. For example if a beam of light could be seen throughout the experiment there will always have been billions of photons in the system. So, even after the reduction in the strength of the light source the 'hits' the experimenters recorded were almost certainly still due to so called stray photons.

Dimming the light from an artificial source, such as a domestic light bulb, normally involves reducing the voltage that is being applied across the bulb's filament. This reduces the level of energy in the system and, consequently, the total amount of energy in the photons that are being emitted by the bulb. But it does not significantly reduce the quantity of photons. There are many everyday sets of circumstances that confirm this. Even in fairly poor light most people can still read the very small print of a newspaper. The only real difference in poor light, as opposed to good light, is that the contrast between the black letters and the white of the paper is greatly reduced (4). This is because in dim light the brightness of the white paper has been greatly lowered. Normally, we compensate for this lack of contrast by holding the newspaper closer than we normally would. By doing this a poorly lit newspaper can be comfortably read, and this is because the outline of every letter of every word is still fairly sharp. Moving closer to the paper would not work unless the difficulty we were having was at least partly due to weakness of reflected light. Moving closer would not help at all if the problem were solely a sparsity of photons. This suggests that the dimming of a source of light does not significantly reduce the number of photons that are being emitted and ultimately being reflected from objects. But, light is never that simple.

There are undoubtedly fewer photons entering our eyes when we describe light as dim than there are when we describe it as bright. But can the difference between poor and bright light be explained solely in terms of numbers of photons? Ultimately, the brightness of the world about us at any given time is determined by the strength of the signal that is entering the visual cortex from the receptors at the back of the eye. Is this strength in turn determined only by the quantity of photons striking the retina? It is dangerous to look directly at the sun. To do so can cause blindness. And, here, intuitively this seems more to do with intensity than simply numbers of photons. When a light is brightened it seems that not only are there more photons being emitted from the source but also that after the adjustment the photons being emitted pack more punch than the photons that were being emitted prior to the change. Photons are not particles. They are probably better described and understood as tiny points of energy. Simply, light is all about sight, and there is no way the eyes could cope with billions upon billions of photons as particles entering them every second. The eyes would be jam-packed full of particles within less than a second of exposure to normal light. Attempting to solve this problem, many theorists argue that, on striking a surface they can not pass through, photons as particles dissipate as heat. This can not be correct. The level of heat that would be generated by so many billions of particles would undoubtedly be painful and damaging. Anyway, matter can not be destroyed through dissipation, or any other means. If photons were particles the filament of a domestic light bulb would burn out within a few seconds. It seems that the energy level of a photon can vary. Maybe photons come in different sizes. Our experiencing near-darkness probably has more to do with a reduction in energy than it has with a photon count.

We know little for sure about light. We can be fairly certain, though, that photons never come in small numbers. Probably the only way to ensure that photons do not enter a particular space is to physically block them. If there are photons around at all they are undoubtedly in their billions.

Basically, it is the pattern of the pulses of photons in the array of light that is reflected to our retinas that gives the brain information about the size, shape and position of objects in the world about us (7). But, it is the quantity of photons that enables us to pick out the very fine details and texture of objects. The strength of the pulses of photons determines the brightness we see, and the frequency of the pulses give rise within the visual cortex to the phenomenon of colour (4). The image that is formed on our retinas by the reflection of photons from an object such as a newspaper could be as small as 2 or 3 mm square, and this is the paper that our brain actually reads. The words on the pages of a retinal image of a newspaper are infinitesimal. Our seeing anything reasonably well involves the transmission of a sharp image and a strong signal between our retinas and the visual cortex. So, if we wish to read a newspaper, especially in poor light, the photons being reflected from its surface have to be in such quantities as to produce a sharp enough image on our retinas. There is very little if any difference in the sharpness of retinal images that are created in good and poor light. In moonlight we may be able to read only the headlines of a newspaper but this still requires the reflection of billions of streams of billions of photons.

In the test of Entanglement theory described above it seemed that the experimenter's joy in finding and eliminating what they believed was the only possible source of false results influenced their perception such that trials that they had previously been writing off as false positives were now clear-cut examples of teleportation. And, it seems that this sort of mindset is at work in determining the outcome of all quantum experiments. The validity of the claims that quantum researchers make about the extraordinary properties that particles have are entirely dependent on whether experimenters really can isolate and control individual photons. It seems that quantum physicists have no doubts at all that photons can be individually accessed, but they are wrong. Nobody understands electromagnetic energy or its constituent parts, photons. The dimensions involved are beyond human grasp. In normal light, a million billion photons are reflected from the head of a pin every second (5). Einstein calculated that photons have no mass and described them as discreet packets of energy or quanta. Mistakenly, though, he believed that they gained mass as they travelled. Light as particles is incompatible with sight. In the Entanglement experiments we have to accept that something happened to something at the near location, and that the same something happened at the distant location but, here, rather than guess that magic was at work, the logical conclusion to come to is that similar effects were produced by similar causes. That is, streams of photons.

Travelling at 300,000 km per second and pulsing with a wavelength of between 400-700 nanometres, in the visible range, light provides the billions of photons the brain needs to construct a picture of the outside world. Seeing involves the reflection of countless billions of photons coming from every part of the world about us and entering our eyes. As we look at an object (say) about a foot high, and about a foot away, light is being reflected from every tiny part of that object and going in every possible direction. Amongst these streams some will be going from the top and from the bottom of the object towards, and converging at, our eyes. For most intents and purposes, the number of streams of photons streaming towards our eyes within this imaginary cone can be regarded as infinite (7). (Could unlike poles be at work here – drawing the entire visual array, but perhaps especially the streams of photons within the aforementioned cone, towards our eyes? Are our visual receptors acting as visual aerials?) In this context, the notion held by many physicists and psychologists of photons as particles, is clearly a preposterous one. They argue that on striking a surface the photon, as a particle, dissipates as heat. Newton established that matter cannot be destroyed and this has never been refuted. Matter cannot be dissolved into nothing. And, as explained earlier, an eye simply could not cope with the heat that countless billions of particles arriving every second would generate.

Tests of Quantum theory normally require the creation and use of single photons. This immediately raises the question of how this is achieved. Light is generated by chemical reaction, as on the sun, by burning combustible materials and by forcing a stream of electrons, a current, through a resistor. The source of experimental light in a laboratory is normally a small, very fine filament in a lamp or a laser. And, at the point in experiments where single photons are needed reports mention the control of their source of light simply being turned down to a very low level. Is this credible? There are a million atoms in the cross sectional area of a human hair (5). So, even the finest, shortest possible filament is likely to consist of several billion atoms. Each atom could contain two or more photon-emitting electrons. And, when full power is being applied, each of these electrons could be emitting many photons. When the power supply to an electric lamp is reduced there comes a point when, quite literally, the light goes out. In other words, at one point there are billions of photons being emitted, and at the next there are none. There is no way of controlling just one atom. There is a critical level in the power supply to a lamp at which everything simply shuts down. Frequency reduction, filtering and sampling may be used to reduce the photon output of special lamps, but there is no way of knowing the exact number of photons that emerge from such processes. The device most commonly used to try to monitor the number of photons streaming through or passing a particular point in an experiment is the scintillation counter (5). Here, photons are directed towards a sodium iodide crystal. Each time supposedly a single photon strikes the crystal there is a flash of light. This is greatly magnified and sent to a counter. But this attempt to count photons is naïve and crude. There is no way of knowing how many photons are entering the system, and the production of a flash of light, no matter how small, could be the work of billions of photons. And, it is on this sort of research that the calibration of photon detectors is based. There has to be serious doubt as to the accuracy of counting instruments that purportedly indicate the passage of single photons.

Quantum researchers seem to take little or no account of the fact that their laboratories are jam-packed full of photons, bouncing off every bit of every surface and streaming in every direction. And this light, if not controlled for, is almost bound to distort experimental results and analyses. For a start, in double-slit experiments, for example, which are entirely open, the ambient light makes silly the claim that at certain points in time there are only a few photons in the system. Also, if a light beam can be seen it must be reflecting billions of photons towards the eyes of its beholder every second. But such reflections are streaming in every direction – including towards and through the slits in the double-slit experiment. These reflections may not be playing any part in the creation of the interference pattern on the rear screen in these experiments but they certainly do rubbish the claim that at some point photons are travelling through the system one at a time.

The simple fact that there are patches of light being created on the rear screen throughout the double-slit experiment proves that at every stage there are billions of photons passing through these systems.

Mathematics is about as objective as you can get. And, since there is no room for opinion in science its no accident that maths and science have become bound inextricably. Much of physics can only be done mathematically. But there is a major difference between science and maths. Science has to have relevance in the real world - maths need not. Maths can quite properly go anywhere and do anything it wants to do. Also, in maths the concept of common sense is practically meaningless. In science, ignoring common sense can be perilous.

In the real world, common sense tells us that one object cannot be in two different places at the same time. And, were data to suggest something that defied common sense most researchers would right away check their calculations and apparatus to try to suss where the experiment had gone wrong. They would also immediately assume that there was a gap in their knowledge and understanding of the idea or the material they were testing. In quantum research, however, it seems that even the most outrageous results are acceptable so long as they can be reconciled in maths.

Scientific experiments are all about discovery. And few if any well designed and carried out experiments could ever be regarded as a failure. Both positive and negative results add to our knowledge. When weird or nonsensical results are obtained, most scientists routinely assume that there is a logical explanation, which simply has to be searched for. In the quantum world things are quite different. Here, experiments are not done to try to gain new knowledge. Rather, they are carried out with the aim of confirming firmly held beliefs. Quantum theorists have made it clear that they have no intentions of altering their theories on the nature of photons. To this end, experiments from which unsatisfactory results are obtained do not get mentioned. And, results that are published are invariably vague and incomplete. Quantum researchers seem ever mindful not to give out the sort of rope that their opponents could hang them with. Also, when weird and wonderful experimental results are produced quantum researchers are clearly very willing to come up with equally weird and wonderful explanations. They state for example, that photons alter their behaviour in the light of new information they have gained about another photon, and that the outcome of an experiment can be significantly altered by how much knowledge the researcher has about what is going on at key stages within the experiment. Quantum researchers claim that photons are capable of thinking and have a memory. They state that photons can be aware of their own properties and the nature of their relationship with another photon, but only after and how they have been measured. Prior to this, researchers explain photons sit "in a superposition of all possibilities waiting for the measurement to force it to make up its mind, and consequently, through Entanglement, the mind of its distant partner"(3). This is not science. This is madness. But sanity and normality will undoubtedly return to the study of quanta. It will happen when quantum theorists eventually see the light.

Seven

THE ATOM

Today, the existence of the atom is not questioned. And, it is now almost invariably described in textbooks as resembling a tiny solar system where the nucleus is the sun and electrons are the planets. This is shown in figure 5. It must be kept in mind that there are no lines associated with an atom. The lines in the diagram simply indicate the orbital path of the atom's electrons. The two atoms in the diagram are shown slightly apart, as logically this is what we might expect due to the repulsive force between electrons. All diagrams in this chapter are, of course, nowhere near to scale. An atom is unimaginably small. For example, there are a million atoms in the cross-sectional area of a human hair. Also, there are more than a 1,000 billion billion atoms in a single drop of water. An atom has never been seen and is unlikely ever to be. Also, the dots representing electrons in all diagrams here are many billion times larger than real electrons. Nobody really knows the structure of the atom. All models of the atom are largely speculative, and none is entirely convincing.

Unlike the solar system, where all of the planets, apart from one, orbit in the same plane around the sun's equator, in the conventional model of the atom the orbits of the electrons are equally spaced around the nucleus. But, there is no obvious physical reason as to why electrons should arrange themselves into such neatly displaced orbits. Some would argue that this is caused by the repulsion of similar charges, but as they revolve it is unlikely that electrons are often in close proximity. Why are electrons in orbit around the nucleus at all? Why do electrons whiz round a nucleus at great

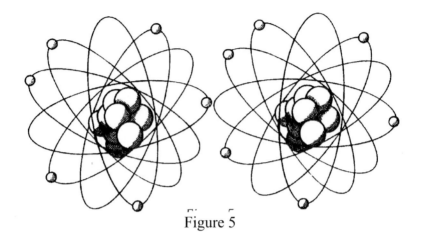

Figure 5

speed? What gave every electron in every atom the kick-start that sent them into perpetual orbit around a nucleus? If the conventional theory is correct every electron in the universe has been orbiting a nucleus for at least 16 billion years.

According to the conventional model every atom contains protons, neutrons and electrons in equal numbers. Protons combine with neutrons and other protons to form the nucleus of the atom. Electrons always function individually. The strong nuclear force facilitates the formation of the nucleus by overcoming the repulsive force that exists between protons. It is postulated that the equality in the number of the different particles gives the atom balance. However, in view of the huge difference in size between protons and electrons this has to be seriously doubted. A proton is nearly 2,000 times bigger than an electron, which means that in heavier elements an electron can be 100,000 times smaller than the nucleus. This surely seriously challenges the conventional claim that atoms are in balance simply because they contain equal numbers of protons, neutrons and electrons.

In the conventional model the number of electrons in each orbit increases with distance from the nucleus. So, depending on the atom in question, there are just 2 electrons in the innermost orbit, 4 in the second orbit, 8 in the third and so on up to 32 or more in the furthest away orbit. This is entirely at odds with the fact that the strength of a magnetic field weakens as the distance from the magnet increases. Since the magnetic attraction is much greater nearer the nucleus, why are there fewer atoms in the closer orbits than there are in the wider ones? Also, why do the electrons in the inner orbit not crash into the nucleus? Furthermore, how could a magnetic field determine precisely the number of electrons in each orbit?

As we know, atoms are birds of a feather. However, current models of the atom do not explain why they do stick together. Rather, they suggest that adjacent atoms would in fact be repelling each other. This could be due to the simple repulsion between just two electrons in a gas or to the comparatively huge repulsive power created by the many orbiting electrons of an atom of one of the heavy elements in close proximity to the orbiting electrons of another such atom. Most theorists believe that there is a weak nuclear force and it is this that holds together the atoms of an element or indeed a piece of any material.

Obviously, a solid is much more difficult to pull apart than a liquid or a gas. This indicates that the atoms of solids are much more tightly bound together than those of gasses. However, there is nothing about either the conventional or the quantum model that indicates how this stronger binding is achieved. The conventional and quantum models are in fact suggesting that the atoms of solids are no more tightly bound than those of gasses. The diagram at figure 6 shows cross sections of neighbouring hydrogen atoms. The pair of atoms at figure 7 represents the much heavier atoms of mercury. These sort of diagrams probably better represent the quantum position, which is that the electrons in an atom do not actually orbit the nucleus. Rather, they are held in 'shells' around the nucleus. Quantum theorists do state, however, that it may be useful to think of electrons as being in orbit. These diagrams show that, according to current theories, the atoms of heavier elements are probably bigger than those of lighter elements.

Any reasonably acceptable model of the atom must at least not have any of the weaknesses mentioned above, that are found in both conventional and the quantum models. Essentially, the design of a good alternative to the design of the existing models of the atom must be consistent with the behaviour of elements that we see in both tests and in everyday life. As we know, the atoms of elements other than gasses like to stick together. And, as we have seen, current models of the atom do not satisfy this. Common sense suggests that the attractive force between the different polarities within an atom would be involved in holding things together. A balanced atom could be put together by taking all of the electrons associated with a particular element and placing them as a sphere around the nucleus. See figure 8, where the atom on the left represents a gaseous atom and the atom on the right a metallic one. In this set-up the pulling power of the nucleus, the repulsive force of the electrons and the number of electrons will, together, determine the size of the atom. Here, the nucleus will be constantly trying to pull all electrons towards it. However, the electrons can only be drawn in as far as the repulsive power between them will allow. At some point, a balance of the forces within this design of atom will be reached. In this model, the electrons of the atoms of lighter elements are more loosely arranged, that is, they are further out from the nucleus and further apart. In other words, the atoms of lighter elements are more spacious than those of heavier elements. So, here, the number of electrons in the atoms of a particular element is determined by the size and strength of the nucleus of the atoms of that element. But, this may well not be, and need not be, as precise as suggested in current theories. Electrons are so tiny in comparison to any nucleus it is surely most unlikely that a nucleus could control electron numbers with exactness. In any case, absolute precision is not required. The idea that an atom is in balance when the actual number of protons and electrons are exactly equal is absurd.

Current theories suggest that it is the weak nuclear force that holds elements together but do not explain how this works. However, the deep field of negatively charged electrons that presumably surround a nucleus suggests that neighbouring atoms would be tending to push each other apart. In our alternative model all electrons of an atom are arranged as a single layer. This could allow the attractive force of a nucleus, which undoubtedly extends beyond its own electrons, to interact with the electrons of adjacent atoms. In the case of solids this could mean that atoms could be drawing each other together to an extent that they have one or more electrons in common. This is illustrated in figure 9. Here, every nucleus is attracting at least one electron belonging to a neighbouring atom as strongly as it is attracting its own electrons. In gasses, perhaps, this common attraction does not exist. What about free electrons? The diagram at figure 9a is a detail of the area in figure 9 where the arrangement of the atoms is such that a small triangular gap exists between adjacent atoms. It is proposed that it is in this gap that the free electrons involved in electric current will normally reside and, of course, travel through when a voltage is applied. Figure 9 could well be a cross section of part of an electric cable.

Despite the popularity of the atom as a nucleus surrounded by electrons nobody has yet explained how the like-charged protons of a nucleus stick together. Theorists can only suggest that there is also a strong nuclear force that overcomes the electromagnetic repulsion. Positively charged particles must repel other positively charged particles, and yet, for most intents and purposes, the nucleus of an atom may be regarded as a single piece of matter. However, until we know better, we may simply have to accept the strong nuclear force as the answer to this anomaly.

Most elements can be easily identified one from another. There are gaseous and metallic elements, and among the metals, gold, copper and iron, for example, are obviously very different from each other. Theorists maintain that these differences are simply due to a difference in the number of protons, neutrons and electrons in the atoms of a particular element, that all particles are matter and that all matter is exactly the same. But, could just an additional handful of protons, neutrons and electrons change an element from a gas to a solid?

At nearly 2,000 times smaller than each proton there is surely doubt as to what extent electrons can determine the characteristics of an atom. The electrons in the atoms of a fairly heavy element could be more than 100,000 times smaller than the nucleus. And even in a very heavy atom there may only be 80 electrons in total. An atom is almost entirely empty space. Also, the nucleus constitutes 99 percent of the mass of an atom. So, it is surely the nucleus of the atom that determines how a particular element looks, feels and behaves. But even if this is the case, the question still remains as to whether numbers alone could make one element quite different from another. Before considering this let us try to appreciate the dimensions within an atom.

If the nucleus of a hydrogen atom, comprising one proton and one neutron, was the size of a golf ball the sole electron of this atom would be orbiting 1km away from the nucleus, making the atom 2km in diameter. Alternatively, if the nucleus were the size of a marble the extent of the atom would be the length of a football pitch. Now, according to current thinking, adding just a few golf balls and the same number of electrons would change considerably the properties of this imaginary atom. Note that in this model electrons are about the size of a pinhead. For example, also in the context of our model, adding just two golf balls to the nucleus of a neon atom comprising ten balls changes this atom to a magnesium atom. Does this ring true? Could a tiny alteration to a relatively small cluster of spheres of the same material bring about as dramatic a change as existing theory suggests. Keep in mind the fact that in our model the golf balls represent the nucleus of an imaginary atom, which is bounded by a few pin-head-sized balls in a 2km diameter empty space, and whose nearest neighbouring nuclei are more than 2km away. All the golf balls in our model are of course exactly the same. So, effectively all we are doing by adding two balls to ten balls is slightly increasing the size of what is effectively a single piece of material. In the real world, particle theorists maintain that all particles are the same in that they are bits of matter, and that, consequently and ultimately, everything that exists is made of matter. In our model, does it seem likely that such a small change could significantly alter the nature and properties of the atom and ultimately the element in question? How could adding just two protons to what is effectively a single piece of matter alter entirely the characteristics of an atom and ultimately an element? Is it not more likely that the protons in the atoms of element X are made of substance X and that the protons in the atoms of element Y are made of substance Y? Could there actually be more than 90 quite different substances in the Universe?

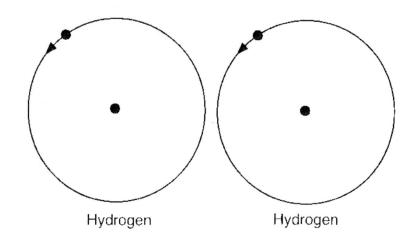

Hydrogen   Hydrogen

Figure 6

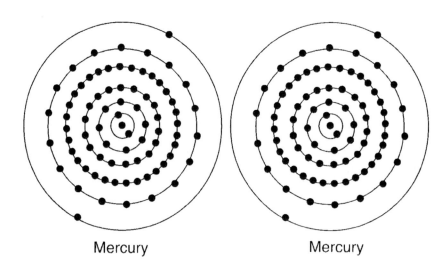

Mercury   Mercury

Figure 7

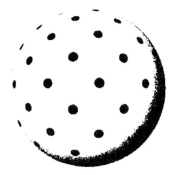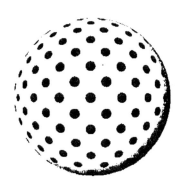

Figure 8

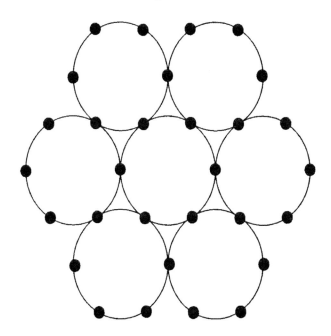

Figure 9

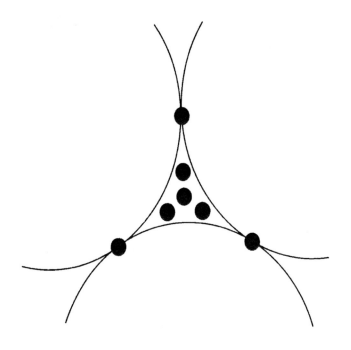

Figure 9a

REFERENCES

(1) PENROSE R, *The Emperor's New Mind*, Vintage, 1990.

(2) ACZEL A D, *Entanglement*, John Wiley & Sons Ltd, 2003.

(3) AL-KHALILI J, *Quantum*, Weidenfeld & Nicolson, 2003.

(4) GREGORY R L, *Eye and Brain*, Oxford University Press, 2004.

(5) *The Hutchinson Dictionary of Science*, Helicon Publishing Ltd, 1993

(6) JENKINS F A and WHITE H E, *Fundamentals of Optics*, McGraw-Hill, 1981.

(7) HABER R N and HERSHENSON M, *The Psychology of Visual Perception*, Holt, Rinehart and Winston, 1973